Psilocybin Mushrooms
of The United States:

A Visual Guide

Publisher's Cataloging-In-Publication Data

Beck, Mary, author
Psilocybin Mushrooms of the United States: A Visual Guide / Mary Beck.

Paperback ISBN-13: 978-1-953450-99-9
Hardback ISBN-13: 978-1-953450-61-6
Ebook   ISBN-13: 978-1-953450-62-3

Library of Congress Control Number: 2021945057

1. Nature—Plants—Mushrooms. 2. Science—Life Sciences—Mycology. 3. Mathematics and Science—Biology, Life Sciences—Mycology, Fungi. I. Mary Beck. II. Title. III. Title : A Visual Guide.

NAT022000 / SCI094000 / PSQ

Mockingbird Press, Augusta, GA
info@mockingbirdpress.com

MARY BECK

Psilocybin Mushrooms *of* The United States:

A Visual Guide

MOCKINGBIRD
—— PRESS ——

BR – © 2013 weed lady (Sylvia); 502230

27 L – © 2013 Richard Kneal (bloodworm); 302350
TR – © 2013 Benjamin Dion (benjamindion); 338496
BR – © 2014 Stephen (Ιερονυμοσ); 425349

28 L – © 2009 Lipa; 43633
TR – © 2014 Django Grootmyers (dgrootmyers@gmail.com); 428854; "IMG_1643.JPG"
BR – © 2011 Panz; 185858

30 TL – © 2013 Benjamin Dion (benjamindion); 326626
TR – © 2013 Richard Kneal (bloodworm); 302351
BL – © 2010 Blue Helix; 83012
BR – © 2013 Richard Kneal (bloodworm); 302352

32 TL – © 2013 Benjamin Dion (benjamindion); 326787
TR – © 2019 Valentina Ramírez (odradekish); 1110504
BL – © 2018 Excited delirium [EXD] ▾ (Excited delirium [EXD] ▾); 884547
BR – © 2013 Benjamin Dion (benjamindion); 338496

34 TL – © 2020 T727; 1190803
TR – © 2017 Rudy Diaz (rdiaz); 744300
BL – © 2019 Chris Mckenna (Angry oyster); 1013925
BR – © 2017 Rudy Diaz (rdiaz); 744301

36 TL – © 2017 Ryan Patrick (donjonson420); 775948
TR – © 2011 vjp; 148981
BL – © 2014 Stephen (Ιερονυμοσ); 336398
BR – © 2014 Stephen (Ιερονυμοσ); 425349

38 TL – © 2012 Byrain; 205305
TR – © 2012 NeoSporen; 225699
BL – © 2012 Byrain; 205304
BR – © 2011 Stephen (Ιερονυμοσ); 145651

41 TL – © 2013 Jonathan Salmon (Ganzig); 370185
TR – © 2014 Caleb Brown (Caleb Brown); 461772
B – © 2013 Caleb Brown (Caleb Brown); 346445

42 TL – © 2013 Caleb Brown (Caleb Brown); 342377
TR – © 2013 Caleb Brown (Caleb Brown); 391291
BL – © 2013 Caleb Brown (Caleb Brown); 346444
BR – © 2013 Caleb Brown (Caleb Brown); 346436

44 TL – © 2014 Caleb Brown (Caleb Brown); 461770
TR – © 2014 Caleb Brown (Caleb Brown); 470509
BL – © 2014 Caleb Brown (Caleb Brown); 461773
BR – © 2014 Caleb Brown (Caleb Brown); 461772

47 T – © 2013 Rocky Houghtby; 460037
BL – © 2010 Lipa; 80206
BR – © 2019 Chris Cassidy (cmcassidy); 1057843

48 TL – © 2016 Terri Clements/Donna Fulton (pinonbistro); 643812
TR – © 2013 Rocky Houghtby; 365724
BL – © 2016 Terri Clements/Donna Fulton (pinonbistro); 643815
BR – © 2013 Rocky Houghtby; 365723

50 TL – © 2012 Evan Casey (EvanCasey); 213030
TR – © 2010 Ron Pastorino (Ronpast); 81514
BL – © 2010 Lipa; 80207
BR – © 2012 Evan Casey (EvanCasey); 213027

53 L – © 2015 Anglerfish; 527875
TR – © 2010 Tom (LanLord); 123216
BR – © 2016 CaraCrass; 651770

54 TL – © 2012 Randy Longnecker (Randy L.); 293319
TR – © 2010 Tom (Lanlord); 123216
BL – © 2010 Tom (Lanlord); 123059
BR – © 2012 Randy Longnecker (Randy L.); 293317

56 TL – © 2012 A. Cortés-Pérez (Alonso); 267585
BL – © 2012 Wolfedawwg; 272332
R – © 2018 Tom Preney; 936599

58 TL – © 2017 Caleb Brown (Caleb Brown); 1116716
BL – © 2012 NeoSporen; 218706
R – © 2017 Caleb Brown (Caleb Brown); 1116719

60 TL – © 2012 NeoSporen; 315307
TR – © NeoSporen; 262532
B – © 2019 Shain.stafford@yahoo.com; 1095428

62 L – © 2012 A. Cortés-Pérez (Alonso); 287979
TR – © 2018 Jimmy Craine (doctorghosty); 967569
BR – © 2017 Jimmy Craine (doctorghosty); 729052

64 TL – © 2009 vjp; 50223
TR – © 2016 CaraCrass; 651770
BL – © 2019 Luke Smithson (Mycofreak); 1122669

"The man who comes back through the Door in the Wall will never be quite the same as the man who went out. He will be wiser but less sure, happier but less self-satisfied, humbler in acknowledging his ignorance yet better equipped to understand the relationship of words to things, of systematic reasoning to the unfathomable mystery which he tries, forever vainly, to comprehend."

-Aldous Huxley

TABLE OF CONTENTS

WARNING

Death and serious bodily injury can occur from the consumption of wild mushrooms. Several wild mushroom species common to the United States, including some identified in this book, contain toxic chemicals dangerous to humans and animals. Death, liver failure, and serious illness have occurred after the consumption of such species. The Publisher and Author do not intend for this book to promote the consumption of wild mushrooms or to serve as a guide for the identification of mushrooms safe for human consumption.

Visual identification of wild mushrooms for purposes of consumption is inherently dangerous. Trained mycologists will often disagree on the identification of mushroom species through unaided visual inspection. Microscopic visualization of mushrooms is required for an accurate identification of a mushroom species. Even with the aid of microscopy, professional mycologists may disagree as to the identity of a mushroom. Amateur mycologists or mushroom foragers should never solely rely on field guides, descriptions, or visual images to identify mushroom species, especially those intended for consumption. The images contained in this book were authored by amateur mycologists and have not been peer reviewed by any professional mycologist or scientific organization. The source of the images is provided. Readers should independently investigate the source and its accuracy.

Psilocybin is a psychoactive chemical that may cause serious harm to humans. Dangerous hallucinations, anxiety attacks, and psychiatric disorders as a result of psilocybin consumption have been reported. The Publisher and Author do not intend for this book to promote the consumption of psilocybin or mushrooms containing psilocybin.

Author & Publisher disclaim any guarantee or warranty of accuracy for contents of this book. The written contents of this book and its graphics are based on published information from various sources, some peer reviewed, some not. The sources for the contents of this book are cited in footnotes and in front and backmatter. Neither the Author nor Publisher guarantee or warranty the accuracy of the information.

Possession of psilocybin may be illegal in the United States.
The Publisher and Author do not intend for this book to promote the possession of psilocybin or mushrooms containing psilocybin. This book does not intend to advise on the legality of possessing or consuming psilocybin.

Author & publisher have not investigated the safety of any species of mushroom for consumption by humans.
The inclusion of any mushroom species in this book is not intended to indicate that the species of mushroom is safe for human consumption. Nothing in this book should be interpreted as indicating a species of mushroom is safe for human consumption.

This book should not be relied upon as a basis for accurate identification of any species of mushroom. Accurate identification of any species of mushroom requires the aid of a trained and professional mycologist with the aid of microscopy.

EDITOR'S NOTE

While mycological topics are discussed in some entries, this book is not intended to be a scientific work. Rather, its purpose is to provide a quick visual reference for readers interested in psilocybin mushrooms of the United States. The book's layout and format were selected in order to achieve that intent.

The reader will find each entry's focus is the images, graphics, and chief characteristics of the species. To avoid larding the text with references and distracting from that purpose, I have decided to place footnotes of the major sources used in a simple, non-scientific format instead of the typical in-line citations common to biology texts. If the reader is interested in more information in an academic form, we suggest they seek out the references listed in the footnotes. I also chose to forgo an index, as it would add little value to a quick reference guide, which this book is intended to be.

A note regarding the bulleted characteristics of the species: Mushrooms will display different characteristics depending on their stage of development. Young mushrooms will exhibit different colors, shapes, ring types, etc., than a fully mature or spent mushroom. The characteristics used in the bullets reference the species at its maturity. Further explanation of characteristics displayed at different stages of growth will be found in the text.

A note regarding claims of potency: all claims contained in this book are anecdotal and obtained from various sources, some non-scientific. There have been scientific studies on potency and attempts to standardize a scale for comparing different species' content of psilocybin, psilocin, and baeocystin. However, widespread analysis has not been conducted and considerable variations in chemical levels can be seen in the same species depending on environment and other factors. For these reasons, the claims of potency in this book should not be taken as authoritative.

Similarly, the maps included in this book are based on anecdotal observations found on www.mushroomobserver.org. These observations are generally from amateur mycologists and lack peer review. A description of a species' geographical range may be found in the textual description.

I hope the choices we made enhance the readers' experience and make this book useful in the pursuit of their mycological explorations.

-**Mary Beck**

INTRODUCTION

Psilocybin mushrooms—also known as "magic mushrooms," or "psychedelic mushrooms"—have been prized by mankind for thousands of years. Unlike LSD, or other synthetic drugs, psilocybin is a naturally occurring chemical present in hundreds of varieties of mushrooms around the world. Botanists classify psilocybin as an "entheogen," or natural substance which has mind-altering properties when ingested. Many anthropologists believe entheogens played a major role in the development of human culture. The ethnobotanist Terence McKenna argued that, in fact, the mind-expanding properties in psilocybin helped create early human society. The chemical appears to act on our brains' language-forming centers and, McKenna believed, the mushrooms may have encouraged mankind's development of religion, art, and philosophy.

Of course, these claims are difficult to prove and speculative. But what is certain is that human societies have used psychedelic mushrooms for millennia. The earliest records of religious ceremonies involving "mushroom cults" date back to at least 9000 BC in northern Africa (modern day Libya and Algeria). Psilocybin mushrooms were used by humans around the world and revered for their spiritual and healing properties. Today, researchers have taken an interest in learning more about what psilocybin can do to improve our well-being and mental health.

Psilocybin in the Americas

The species Psilocybe mexicana grows naturally in many regions of modern-day Mexico and the United States. It has strong psychedelic properties and played a major role in the spiritual practices of many Native American cultures. The Aztecs named it teonanacatl, which translates roughly to "God's mushrooms" (in English they are commonly known as "liberty caps"). The Aztec god Xōchipilli is usually represented with teonanacatl mushrooms on his knees and his earlobes; his head is often thrown back in apparent ecstasy, and his face carries a trance-like expression.

The early Spanish visitors to the New World documented the Aztecs' use of mushrooms in their religious ceremonies. The Spanish physician Francisco Hernandez de Toledo wrote a guide for missionaries to the New World and included a section on the mushrooms eaten by native peoples. He described—with a certain awe—that some of the mushrooms induced a "madness," causing people to see "all kinds of things, such as wars and the likeness of demons." The Franciscan friar Bernardino de Sahagún also described what he called the "intoxicating sacred mushrooms" that were eaten by the "Indians of Mexico" at feasts and at religious ceremonies.

The Aztecs were not the only culture to revere the Psilocybe mexicana. So-called "mushroom stones" dating back thousands of years have been discovered throughout Guatemala. The stones are carved in the shape of a pileate mushroom, with human figures on the stalks. Anthropologists are uncertain of the stones' significance, but they clearly point to a veneration of the mushrooms. The ceremonial use of Psilocybe mexicana persisted at least into the twentieth century. American investigators in the 1930s reported native groups in southern Mexico still using psychedelic mushrooms. In fact, these anthropologists fed into and helped spread a widening curiosity about the practice.

Psilocybin in modern American culture

In 1957, an American banker named Gordon Wasson took a trip to Mexico. Wasson was curious about native cultures and intrigued after reading about their use of psychedelic mushrooms. Wasson met a tribe still involved in the practice and sent a sample of the mushrooms to Albert Hofmann, the Swiss chemist who discovered LSD. Wasson also published a photo-essay in Time Magazine that year, recounting his experiences with the psychoactive mushrooms.

Wasson's experience, and the Time article, led to an increased interest in psychedelic mushrooms. Just a year after Wasson sent him the sample, Hofmann and his associates managed to isolate the compounds which gave the mushroom its psychedelic power (psilocybin and psilocin). This coincided with a general growing fascination with psychedelic drugs in American society, both naturally occurring and synthesized. In 1959, Timothy Leary joined the staff of Harvard University. The following year, Leary and one of his colleagues founded the Harvard Psilocybin Project, which aimed to explore psilocybin's effect on human consciousness. Leary and his team offered psilocybin to volunteers and documented their experiences, which led to further interest in the psychedelic.

Psilocybin in the counterculture

Leary's experiments quickly left the rarified world of Harvard's psychology department and spilled over into bohemian culture. Leary introduced himself to the Beat poet Allen Ginsberg and invited him to try psilocybin. The experiment was a rousing success. Ginsberg apparently ingested the psilocybin in an upstairs room; several hours later he came down the stairs stark naked and proudly announced that he was the Messiah. He became an enthusiastic psilocybin user and introduced his friends to the drug.

Around the time that Ginsberg was sampling psychedelic mushrooms, his old friend William Burroughs was also experimenting with entheogenic plants. Burroughs was, of course, justly famous for his drug use—much of which he documented in *Junkie* and *Naked Lunch*. He had already sampled peyote on a trip to Mexico, but disliked the experiment. "Everything I saw looked like a peyote plant," he said of the trip. Undeterred, he set out on an expedition around South America, looking for ayahuasca, a psychedelic tea brewed from the Psychotria viridis shrub. During his search, Burroughs went to Bogota University, where he met a botanist and expert on hallucinogenic drugs named Richard Schultes. Schultes took Burroughs on an expedition down the Amazon, eventually finding and sampling ayahuasca.

Burroughs declared that ayahuasca was everything he had hoped it would be. However, psilocybin was a disappointment for him. In 1961, Timothy Leary met up with a group of Beat writers in Tangier—the gathering included Burroughs, Ginsberg, Paul Bowles, and Gregory Corso. Leary brought psilocybin pills to the party and almost everyone enjoyed the experience. Burroughs, however, did not. Leary later wrote that Burroughs, in a state of near-collapse, said, "I would like to sound a word of warning. I'm not feeling too well. I was struck by juxtaposition of purple fire mushroomed from the Pain Banks. Urgent Warning. I think I'll stay here in shriveling envelopes of larval flesh."

Ultimately, the surge of interest in psychedelics may have also signaled a break between the era of the Beat Generation and the hippies. Ginsberg and a few others were able to fit in with the burgeoning psychedelic culture, but others, notably Jack Kerouac and William Burroughs, simply were not. The hippie movement, though, embraced "magic mushrooms" whole-heartedly. They were seen as a pathway to enlightenment and spiritual experience—a way to bypass social conditioning and reach for the stars.

Legal issues

Psychedelic mushrooms were legal in the United States until 1970. Leary's experiments may have raised a few eyebrows, but they did not land him in any legal trouble. However, in 1970, psilocybin was classified as a controlled substance by the federal government. It is currently listed on Schedule I of the Controlled Substances Act, a category restricted to heavily criminalized drugs which are considered to have a high potential for abuse. That may explain why relatively few people say they have taken psilocybin. In a 2010 survey, only 0.1% of respondents said that they had used any form of psychedelics in the past year. Because of its classification, scientists and researchers must go through an arduous, expensive application process to carry out studies of psilocybin's properties. The days of Timothy Leary's free-wheeling experiments were long gone indeed.

In recent years, though, there have been signs that the public attitude toward psychedelic mushrooms is shifting. Some states and cities have moved toward decriminalizing psilocybin. Denver, Colorado, did so in 2019, making it illegal for the city government to use resources to impose criminal penalties against adults for using or possessing psilocybin. (The substance was not, strictly speaking, made legal, but the law removed the fear of criminal prosecution—it did not, however, allow mushrooms to be bought or sold.) Similar laws passed in Santa Cruz and Oakland, California, and there is a movement to pass legislation to this effect in Washington D.C. A lawmaker in New York state recently introduced legislation that would decriminalize psilocybin throughout that state. There are also plans to make psilocybin legal for therapeutic purposes in Oregon.

Properties of Psilocybin

Psilocybin is the name for the chemical compound which occurs naturally in many species of wild mushrooms. Upon ingestion, the body almost immediately converts psilocybin to a substance known as psilocin. Psilocin is a mind-altering substance which may be mildly hallucinogenic. Its effects will vary greatly from person to person, and even from day to day. Broadly speaking, though, the experience of taking psilocybin is comparable to the experience of taking LSD or mescaline.

The "high" from psilocybin normally lasts between four and six hours, with a peak of intensity about two or three hours after the substance is ingested. (The manner of ingestion can vary greatly too. Participants in scientific studies will likely take psilocybin in a capsule form. Other users may eat pieces of dried mushroom, or drink a tea made from the dried mushrooms.)

Psilocybin leads to an overall sensory enhancement—colors look bright and saturated, while sounds are rich. Some users report a heightened sensitivity to touch and smell, and some shifting of the boundaries between the senses, so that they may "smell" colors or "taste" sounds. In general, users' vision takes on a certain flow. Objects may gleam around the edges and lines may be blurry. In some cases, users report mild hallucinations, most often seeing geometric shapes overlaid on other objects. The most striking features of the psilocybin experience, though, is a highly charged mood and a greater than average sense of openness to new, hitherto unexplored ways of thinking and a heightened sense of personal awareness.

Psilocybin and mental health

Researchers have been excited about psilocybin's possibilities for decades, precisely because of the kind of openness that the drug engenders in its users. Psilocybin seems to give users new insight into their own lives and into their mental processes, so that they are able to step away from bad patterns and re-imagine their lives in new ways. Psychiatrists have suggested that the drug could be invaluable in treating depression, anxiety, and opium addiction, among other things. To date, a number of studies have been carried out on psilocybin's potential to treat depression and anxiety and the results appear promising.

Robin Carhart-Harris, a researcher at Imperial College London, conducted one such study and reported a sense of awe when he supervised the patients who were undergoing treatment. "It is almost like being in the presence of someone particularly wise, in terms of what comes out of their mouth," he marveled. Patients have reported that their depression lifted after a single dose of psilocybin, leaving them feeling calm and happy for months on end.

As of yet, psilocybin is a highly restricted substance, which makes it materially difficult to carry out tests into its possible uses. While testing is not prohibited, the process of applying for permission is lengthy and expensive. This has made it difficult to really assess psilocybin's potential as a treatment for patients in the long-term. This is also why many scientists are expressing great optimism about signs that the U.S. is becoming more tolerant of psilocybin. As more locations decriminalize the drug, it will likely get easier to conduct tests of its potential.

For now, what is known is that psilocybin has very few side effects and appears to have significant benefits. The drug has very low toxicity and is not habit-forming. Some users do experience "bad trips"—panic attacks, a heightened sense of paranoia and dread, and in the worst cases, suicidal thoughts. That is why most experts recommend that one should be accompanied when taking psilocybin. However, even when a psilocybin "trip" does turn negative, those negative effects don't seem to last once the drug wears off. Unlike LSD, or other synthetic psychedelics, mushrooms do not appear to do long-term damage to the human brain or nervous system.

What does this all mean? Perhaps, that humankind was right when, thousands of years ago, they first turned to psilocybin mushrooms as a path to spiritual enlightenment. Perhaps the humble "magic mushroom" is the tool we need today, to put us right with ourselves and with the universe.

SPECIES BY REGION

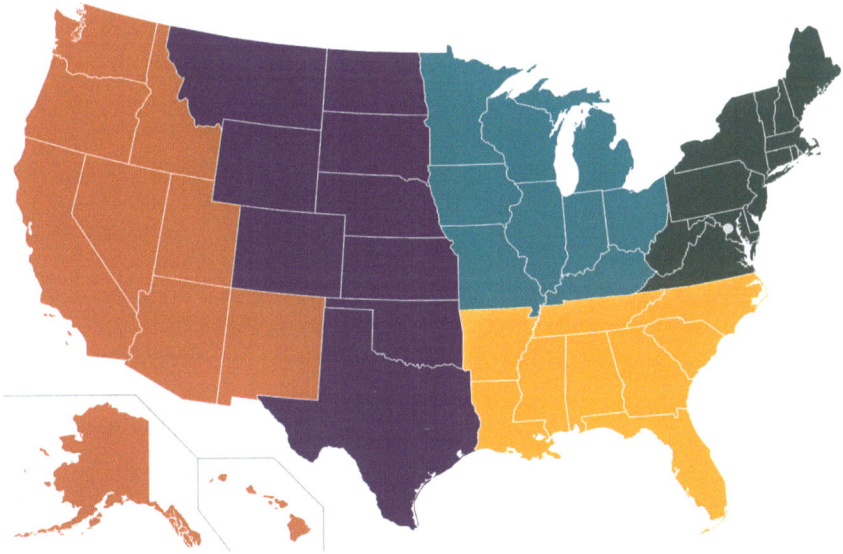

GENUS GYMNOPILUS

The Gymnopilus genus is composed of more than 200 saprobic species of gilled mushrooms. Mycologist Paul Stamets noted that Gymnopilus species are often medium to large compared to other mushrooms, prefer to grow on wood, and have dry caps and prominent veils. The species often have convex caps and adnexed or adnate gills on the hymenium. The stipe may or may not contain a partial ring veil. The mushrooms grow gregariously, in solitude or in scattered groups on decaying wood like stumps, logs, woodchips, and pastures. The spore print of this genus is often a reddish-brown or orange-brown color. Since Gymnopilus species are potent spore producers, the spores can be seen reliably under a microscope, with cheilocystidia always observed on the gill edges.

Many Gymnopilus species have been found to be psychoactive and contain the hallucinogens psilocin and psilocybin. Notable psychoactive species are G. aeruginosus, G. braendlei, G. intermedius, G. luteoviridis, G. liquiritiae, G. luteus, G. purpuratus, G. spectabilis, G. validipes, and G. viridans. In the psychoactive, psilocybin-containing species, the cap flesh and stipe have often been observed to bruise a bluish green color when handled.

Careful observers in the field have noted that Gymnopilus species are most often confused with mushrooms in the genus Pholiota. It is also important to note Gymnopilus is also often misidentified with the poisonous Galerina genus which contains species that can be deadly if ingested. Gymnopilus are usually larger than Galerina. Disconcertingly, however, both have similarly colored spore prints and often show a ring on the mushroom's stipe.

The genus is very well characterized in the scientific literature in terms of morphology. However, a recent study that compared Gymnopilus taxonomy based on morphology to taxonomy based on ribosomal sequence data showed traditional taxonomic classification is not always in agreement with ribosomal sequencing. Specifically, spore dimensions and the absence or presence of a partial veil cannot reliably be used to differentiate species within the Gymnopilus genus accurately. This study created several phylogenetic clades based on ribosomal sequence type, including the groups aeruginosus-luteofolius, lepidotus-subearlei, spectabilis-imperialis, nevadensis-penetrans, and underwoodii-validipes.

References

[1] Guzmán-Dávalos L, Mueller GM, Cifuentes J, Miller AN, Santerre A., Traditional infrageneric classification of Gymnopilus is not supported by ribosomal DNA sequence data, Mycologia, 2003. Nov-Dec;95(6):1204-14. PMID: 21149021

[2] https://www.mushroomexpert.com/gymnopilus.html

[3] https://www.sciencedirect.com/topics/agricultural-and-biological-sciences/gymnopilus

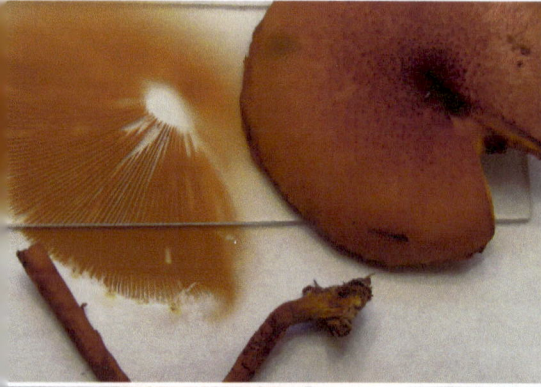

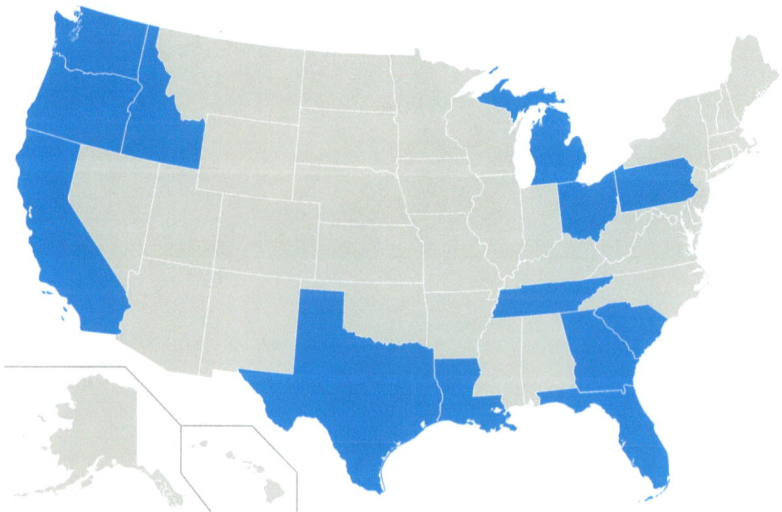

Cap Shape

Convex

Hymenium Shape

Gills

Hymenium Attachment

Adnate or Adnexed

Stipe Character

Ring

Psychoactive

Medium

Ecological Type

Saprotrophic

Spore Print Color

Yellow-Orange

The convex cap of Gymnopilus aeruginosus ranges from 2 to 23 cm in diameter. The cap has been observed to display variations in color, ranging from bluish-gray green to greenish yellow. The cap displays these aforementioned hues when the mushroom is young, and becomes dark brown as the mushroom reaches advanced age. The adnexed to adnate gills of this species are cream-colored to pale orange with irregular edges. The stipe, which has a ring, ranges from 30 to 120 mm in length and 4 to 40 mm in thickness. The stem's coloration is similar to the cap. The spore print color of this species centers around yellow-orange, and can appear slightly rusty orange to rusty brown.

G. aeruginosus grows gregariously or in solitude on decayed conifers and hardwood trees as well. The species has been observed growing on wood chips, sawdust, and tree stumps from spring to autumn. G. aeruginosus is widely-distributed across the United States. The species has been observed in California, Oregon, Washington, Idaho, Michigan, Tennesee, Ohio, and Pennsylvania. It is common in the Pacific Northwest United States, and also found in Japan and Korea.

The mushroom is psychoactive and contains the hallucinogen psilocybin. This species is commonly known as the Magic Blue Gym, its common name, and has been noted to have a bitter taste when consumed. It closely resembles related species G. luteofolius, but the two species can be differentiated by comparing the veil and cap flesh. The G. luteofolius veil is fibrillose-membranous with white cap flesh, whereas the G. aeruginosus veil is just fibrillose and with slightly green cap flesh.

References

[1] Guzmán-Dávalos L, Mueller GM, Cifuentes J, Miller AN, Santerre A., Traditional infrageneric classification of Gymnopilus is not supported by ribosomal DNA sequence data, Mycologia, 2003. Nov-Dec;95(6):1204-14. PMID: 21149021

[2] https://www.shroomery.org/12463/Gymnopilus-aeruginosus

[3] https://mushroomobserver.org/name/show_name/39483

[4] Stamets, Paul (1996). Psilocybin Mushrooms of the World. Berkeley: Ten Speed Press. Pages 179-180. ISBN 0-9610798-0-0.

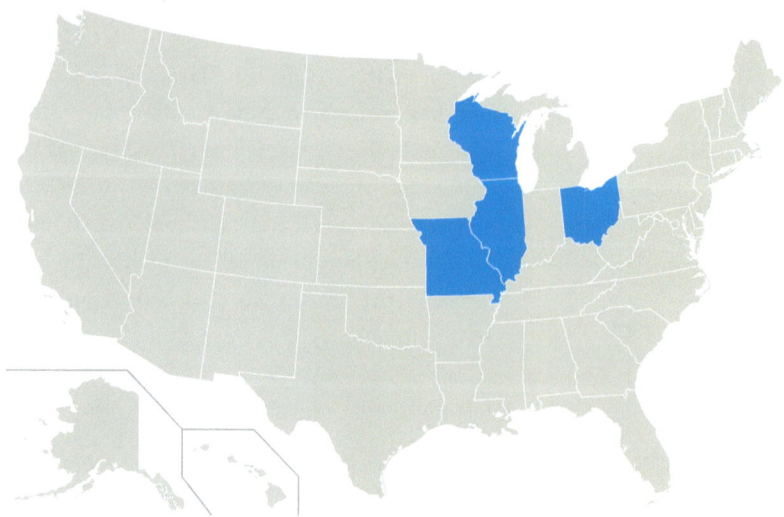

Cap Shape

Convex

Hymenium Shape

Gills

Hymenium Attachment

Adnate or Adnexed

Stipe Character

Cortina

Psychoactive

Unknown

Ecological Type

Saprotrophic

Spore Print Color

Orange-Brown

The cap of the agaric mushroom species Gymnopilus braendlei ranges from 2.5 to 5 centimeters, with a cap shape that is hemispheric and convex. Sometimes, the cap shape can be slightly umbilicate. When the mushroom is young, the cap is a purple hue, turning light pink toward the margins. As G. braendlei ages in its lifecycle, the cap becomes a yellow color with greenish stains. The spore-bearing surface, or hymenium contains broad, close gills which are whitish when the mushroom is young, and turn bright orange-brown with age. The stem ranges from 2.5 to 4 cm in length to 3 to 4 cm in thickness, and tapers from yellow upward from the base, to a pinkish color at the apex. Spores are orange-brown.

G. braendlei can grow in solitude or appear gregariously in small clusters on dead tree stumps throughout its widespread habitat throughout the eastern United States. The species has also been found in the western United States. G. brandlei is purportedly bitter when tasted, and psychoactive, containing the compounds psilocybin and psilocin. It grows in the summer and autumn seasons between June and November. G. braendlei was originally discovered and characterized by prolific mycologist Charles Horton Peck in Washington, D.C. in 1902.

References

[1] https://mushroomobserver.org/name/show_name/18317?q=pDYJ

[2] https://www.shroomery.org/12464/Gymnopilus-braendlei

[3] Gastón Guzmán, John W. Allen, Jochen Gartz (1998). "A worldwide geographical distribution of the neurotropic fungi, an analysis and discussion."

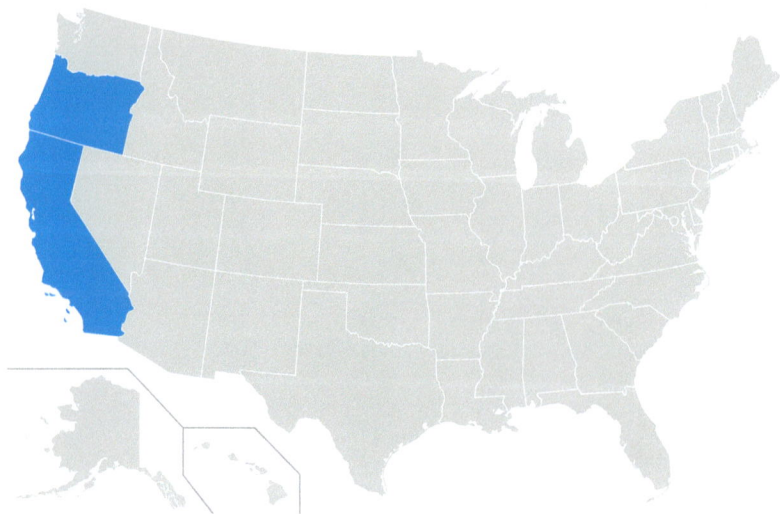

Cap Shape

Convex

Hymenium Shape

Gills

Hymenium Attachment

Adnate

Stipe Character

Ring

Psychoactive

Unknown

Ecological Type

Saprotrophic

Spore Print Color

Brown

The convex to planar convex cap of Gymnopilus dilepis ranges from 2 to 5 cm in diameter. The cap has a light orange surface, with a pale orange center and light orange margin. The adnate to subdecurrent gills of this species range from light orange to orange with dark brown hues appearing toward the base. The stipe ranges from 4 to 8 cm in length and 1 to 2 cm in thickness. The stem has a yellowish flushed color and a partial stem ring veil when the mushroom is young. The spore print color of this species is light brown to yellowish brown.

This is a relatively rare Gymnopilus species that has been observed growing gregariously or in solitude on stumps, dead conifers, or wood chips on the forest floor. It has been observed growing in Britain, Southern Europe, India, North America, and Thailand. In Britain, the common name for this species is the Magenta Rustgill.

It has been noted that this Gymnopilus species is inedible and can be toxic or cause allergic reactions. No published literature on psychoactive properties of this species is readily available.

References

[1] Guzmán-Dávalos L, Mueller GM, Cifuentes J, Miller AN, Santerre A., Traditional infrageneric classification of Gymnopilus is not supported by ribosomal DNA sequence data, Mycologia, 2003. Nov-Dec;95(6):1204-14. PMID: 21149021

[2] https://www.first-nature.com/fungi/gymnopilus-dilepis.php

[3] Suwannarach N, Kumla J, S Kriangsak, Lumyong S, Gymnopilus dilepis, a new record in Thailand. Mycotaxon, 2017. Volume 132, Issue 2, Pages 337-341

[4] https://mushroomobserver.org/name/show_name/20932

GYMNOPILUS JUNONIUS (SPECTABILIS)

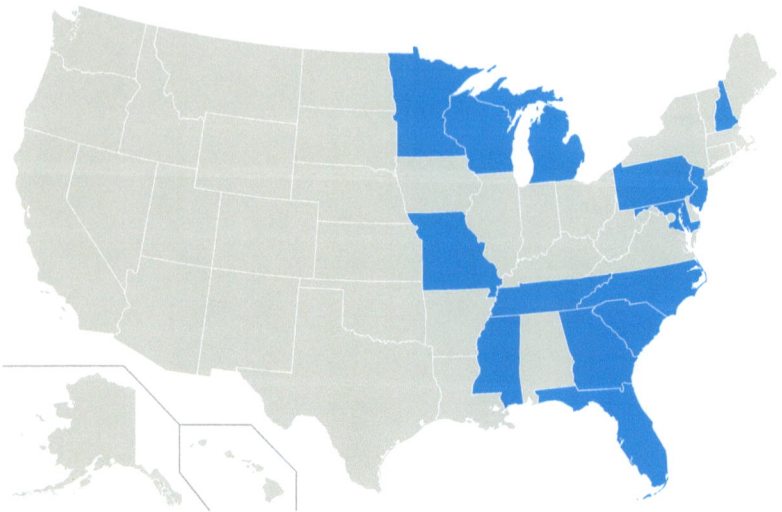

Cap Shape

Convex

Hymenium Shape

Gills

Hymenium Attachment

Adnate

Stipe Character

Ring

Psychoactive

Medium

Ecological Type

Saprotrophic

Spore Print Color

Red-Brown

Gymnopilus junonius is synonymous with Gymnopilus spectabilis. The large convex cap of this species ranges from 5 to 40 cm in diameter and sports a rusty orange to reddish brown color. Like some other Gymnopilus species, the cap bruises blue or green when handled. The adnate gills of G. junonius are pale yellow to rusty orange. The stipe has a ring and ranges from 30 to 250 mm in length and 1 to 10 mm in thickness. The stem is a rusty orange color, like the cap flesh. The spore print color of G. junonius is reddish-brown.

This species grows gregariously or in scattered groups in the summer months through the fall. It has rarely been observed growing in solitude. The species grows on dead conifer trees and decaying hardwood trees. Within the United States, it grows in forests in the Pacific Northwest and on the Eastern Coast. Globally, it is widely-distributed and can be found in South America, Europe, Africa, China, India, Russia, and Australia.

Some subspecies of this mushroom contain the hallucinogen psilocybin, and some subspecies do not. Some subspecies of G. junonius have been found by researchers to contain neurotoxins.

Common names for this mushroom include Big Gym, Giant Laughing Mushroom, Laughing Gym, Laughing Cap, and Spectacular Rustgill.

References

[1] https://www.mushroomexpert.com/gymnopilus_junonius.html

[2] https://www.shroomery.org/12465/Gymnopilus-junonius

[3] Lee IK, Cho SM, Seok SJ, Yun BS: Chemical Constituents of Gymnopilus spectabilis and Their Antioxidant Activity, Mycobiology 2008. Volume 36, Issue 1, Pages 55-59.

[4] Stamets, Paul (1996). Psilocybin Mushrooms of the World. Berkeley: Ten Speed Press. Pages 179-180. ISBN 0-9610798-0-0.

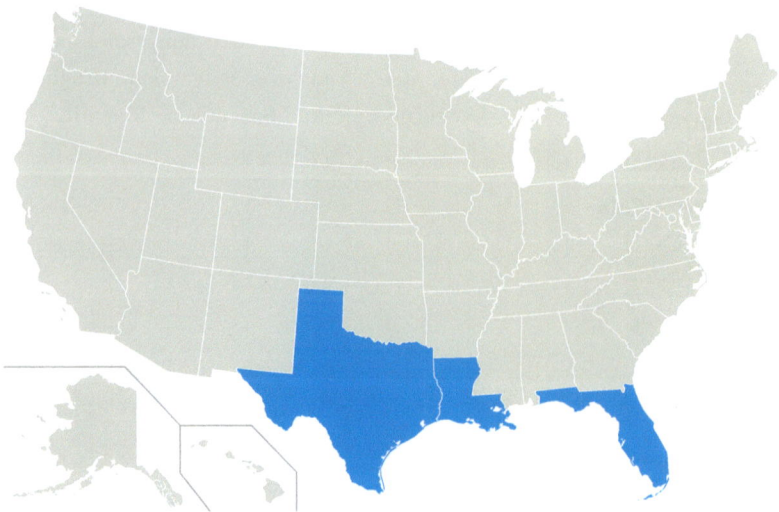

Cap Shape

Convex

Hymenium Shape

Gills

Hymenium Attachment

Adnexed

Stipe Character

Cortina

Psychoactive

Unknown

Ecological Type

Saprotrophic

Spore Print Color

Orange-Brown

The convex cap of Gymnopilus lepidotus ranges from 5 to 8 mm in diameter. The pileus surface is light orange or yellow. The broad, adnexed gills of this mushroom are a rusty orange or yellow. The stipe ranges from 25 to 60 mm in length and 3 to 9 mm in thickness with a pale orange surface that becomes dark brown toward the base. Spores have been observed to be orange-brown to light brown in spore prints.

G. lepidotus has been seen growing gregariously in small clusters. The original specimens described by Hesler and Guzmán-Dávalos were found only in Mexico. More recently, however, most specimens have been observed growing on logs throughout Florida in July.

While psilocybin has not specifically been claimed to be found in G. lepidotus in publications, all mushrooms of the Gymnopilus genus can theoretically produce psilocybin. G. lepidotus may produce an allergic reaction if consumed.

It is sometimes difficult to differentiate between species within the Gymnopilus genus by morphology alone, due to similarities between disparate species and differing morphologies within species. G. lepidotus belongs to the lepidotus-subearlei clade of Gymnopilus, which all contain erect, reddish squamules in the pileus and are restricted to tropical or subtropical environments. G. lepidotus strongly resembles related species Gymnopilus dilepis, found in Thailand, but the two species can be resolved at the microscopic level through differing cheilocystidia.

References

[1] https://mushroomobserver.org/observer/show_observation/136907

[2] https://mushroomobserver.org/observer/observation_search?page=1&pattern=Gymnopilus+lepidotus

[3] Guzmán-Dávalos L, Mueller GM, Cifuentes J, Miller AN, Santerre A., Traditional infrageneric classification of Gymnopilus is not supported by ribosomal DNA sequence data, Mycologia, 2003. Nov-Dec;95(6):1204-14. PMID: 21149021

[4] Hesler LR. (1969). North American Species of Gymnopilus (Mycologia Memoir Series: No 3). Knoxville, Tennessee: Lubrecht & Cramer Ltd. pp. 40–41. ISBN 0-945345-39-9.

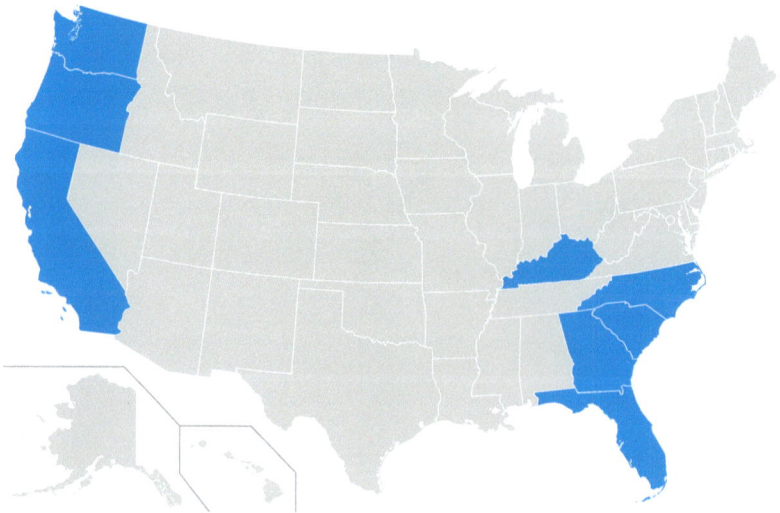

Cap Shape

Convex

Hymenium Shape

Gills

Hymenium Attachment

Adnate

Stipe Character

Ring

Psychoactive

Low

Ecological Type

Saprotrophic

Spore Print Color

Reddish-Brown

The convex cap of Gymnopilus luteofolius ranges from 20 mm to 79 mm in diameter and is commonly dark red to purple-red with fibrillose scales. With age, the cap often fades to a yellow hue, and some observations have noted green staining on the cap. This species has notched to adnate yellow gills that become orange with spores. The stipe, which ranges from purple to yellow in color, ranges from 3 to 9 mm in length to 3 to 10 mm in thickness, and often displays green staining near the enlarged base. G. luteofolius spore prints are reddish-brown or rusty orange, and the mushrooms grow gregariously in clusters.

These clusters are widely distributed throughout both North American coasts. One well-documented collection was additionally described from the Midwest, in Illinois. The clusters grow on the dead wood of deciduous trees and conifers, as well as commercial lumber.

Like some other members of the Gymnopilus genus, G. luteofolius is psychoactive and contains psilocybin. Recently, G. luteofolius has been considered a promising candidate for fungal bioremediation treatment of wastewater to remove pharmaceutical micropollutants.

Phylogenetically, G. luteofolius belongs to the aeruginosus-luteofolius clade of the Gymnopilus genus, including mushrooms with similar morphologies like G. aeruginosus, G. luteofolius, G. cf. punctifolius and G. subpurpuratus. These mushrooms all have blue to green staining that suggests the presence of psilocybin and grow in temperate to tropical climates. G. luteofolius can be differentiated from morphologically-similar species G. aeruginosus by comparing the veil and cap flesh. The G. luteofolius veil is fibrillose-membranous with white cap flesh, whereas the G. aeruginosus veil is just fibrillose and with slightly green cap flesh.

References

[1] Castellet-Rovira F, Lucas D, Villagrasa M, Rodríguez-Mozaz S, Barceló D, Sarrà M, Stropharia rugosoannulata and Gymnopilus luteofolius: Promising fungal species for pharmaceutical biodegradation in contaminated water, J Environ Manage. 2018 Feb 1; 207:396-404. PMID: 29190482

[2] https://www.mushroomexpert.com/gymnopilus_luteofolius.html

[3] http://www.mykoweb.com/CAF/species/Gymnopilus_luteofolius.html

[4] https://mushroomobserver.org/name/show_name/251

[5] https://www.shroomery.org/12467/Gymnopilus-luteofolius

GYMNOPILUS LUTEOVIRIDIS

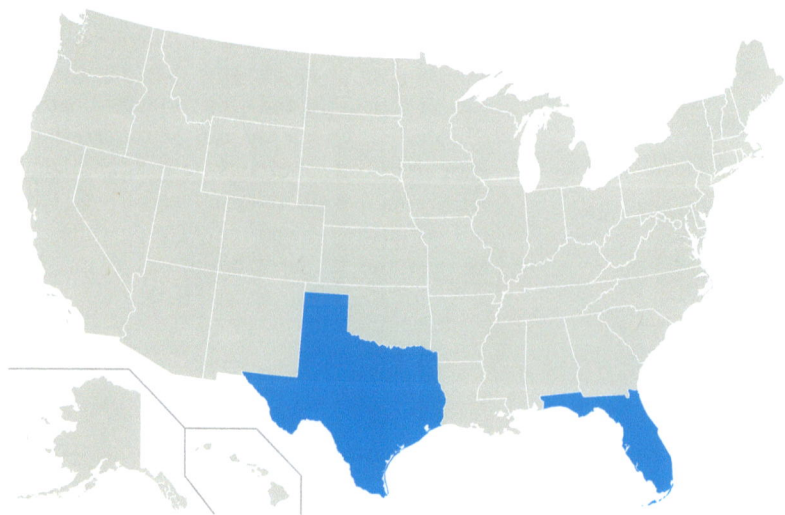

Cap Shape

Convex

Hymenium Shape

Gills

Hymenium Attachment

Adnate or Adnexed

Stipe Character

Ring

Psychoactive

Unknown

Ecological Type

Saprotrophic

Spore Print Color

Yellow-Orange

The convex cap of mushroom species Gymnopilus luteoviridis ranges from 2.5 to 4 cm in diameter and is a straw-colored yellow or mustard yellow. The flesh of the cap is the same color as the cap surface and stains green if injured. The short, decurrent tooth gills are dark yellow and become darker brown as the mushroom ages. The stipe ranges from 4 to 6 cm long by 0.3 to 0.5 cm wide, can be stuffed or hollow, and is cream yellow. The stipe stains green with handling and with age. The stipe contains a partial veil, and the mushroom has a bitter taste if ingested. The spore print of this species is yellow-orange, and it grows gregariously in clusters or dense mats on oak stumps widely throughout Eastern North America.

Like other species within the genus Gymnopilus, G. luteoviridis contains psilocybin and is psychoactive.

Mushrooms within this genus are often difficult to differentiate by morphology alone, as many Gymnopilus species resemble one another. The mushroom was first described by mycologist H.D. Thiers in 1959 in his article, "The Agaric Flora of Texas." In his writings, Thiers describes how G. luteoviridis is best distinguished by its yellow cap and microscopic presence of pleurocystidia.

References

[1] https://mushroomobserver.org/observer/index_observation?by=name&q=vdpk

[2] https://www.shroomery.org/12468/Gymnopilus-luteoviridis

[3.] Thiers, Harry D. (1959). "The Agaric Flora of Texas. III. New taxa of brown- and black-spored agarics." Mycologia. 51 533–535. doi:10.2307/3756141

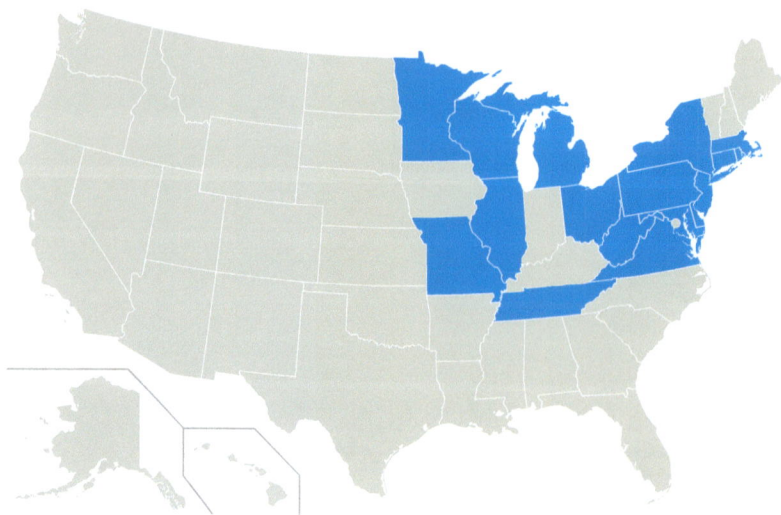

Cap Shape

Convex

Hymenium Shape

Gills

Hymenium Attachment

Adnate or Adnexed

Stipe Character

Ring

Psychoactive

Unknown

Ecological Type

Saprotrophic

Spore Print Color

Reddish-Brown

The agaric mushroom Gymnopilus luteus presents a convex cap that ranges from 5 to 10 cm in diameter. The cap is buff yellow to warm orange, and is often darker toward the middle. When the mushroom is young, the adnexed, thin gills are pale yellow, becoming brownish with age. The stipe ranges from 4 to 8 cm in length and 0.5 to 1.5 centimeters in thickness, containing a partial veil with a subtle ring. The yellow stipe is the same color as the cap and exhibits orange-brown or blue-green stains when handled. G. luteus spore prints are rusty brown.

The mushrooms grow in solitude or gregariously in small clusters throughout Eastern North America on dead hardwoods from June until November.

G. luteus has been shown to contain psychoactive psilocybin in peer-reviewed scientific literature. It is often mistaken for another Gymnopilus species, Gymnopilus junonius. G. luteus can be most easily differentiated from similar species by its yellow cap color and comparatively slender stem. In the book, *Psilocybin Mushrooms of the World*, Paul Stamets writes that Gymnopilus species closely resemble related species in the genus Galerina, which contains poisonous mushrooms if ingested.

References

[1] Hatfield GM, Valdes LJ. The occurrence of psilocybin in Gymnopilus species. Lloydia. 1978 Mar-Apr;41(2):140-4. PMID: 565861

[2] https://www.mushroomexpert.com/gymnopilus_luteus.html

[3] https://www.shroomery.org/12469/Gymnopilus-luteus

[4] Stamets, Paul (1996). Psilocybin Mushrooms of the World. Berkeley: Ten Speed Press. Pages 178 – 185. ISBN 0-9610798-0-0.

[5] https://mushroomobserver.org/observer/observation_search?pattern=Gymnopilus+luteus

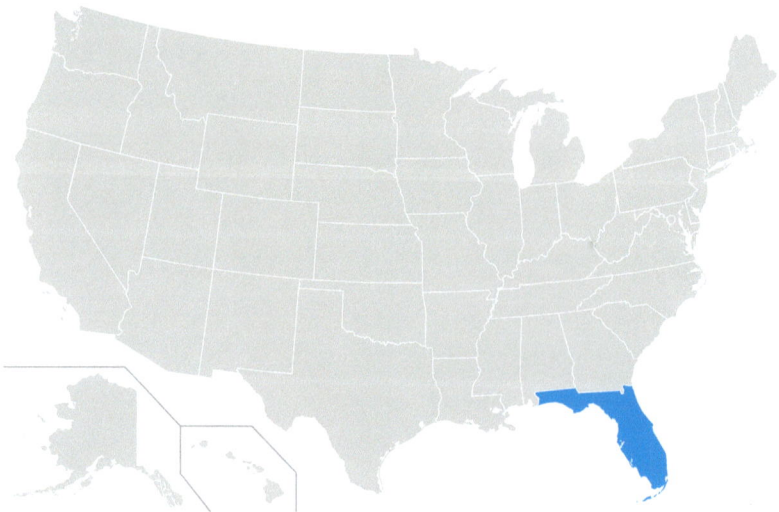

Cap Shape

Convex

Hymenium Shape

Gills

Hymenium Attachment

Adnate or Adnexed

Stipe Character

Ring

Psychoactive

Unknown

Ecological Type

Saprotrophic

Spore Print Color

Reddish-Brown

The relatively small agaric mushroom species Gymnopilus palmicola grows with a convex cap that ranges from 1.5 to 2.5 cm in diameter and has a rusty, brownish orange hue. The broad gills attached to the stem are a rusty brown color, like the cap. The mushroom's stipe ranges from 1 to 2 cm in length and 2 to 3 mm in thickness and is a whitish brown color with a ring near the apex. Spore prints from G. palmicola are a reddish-brown color.

The mushrooms grow in solitude or in small clusters on the logs of dead palm trees and have been observed to grow on living orchids. Like other members of the Gymnopilus genus that are within the aeruginosus-luteofolius phylogenetic clade, G. palmicola grows in tropical to temperate climates from the Gulf Coast to the Caribbean.

It is unclear whether this species contains psychoactive psilocybin. Belonging to the aeruginosus-luteofolius phylogenetic clade suggests it does contain psilocybin, but the lack of blue staining in the species raises doubts. A very similar species, Gymnopilus cyanopalmicola, has been described in Mexico and does stain blue on the stipe when bruised. Mushroom expert Michael Kuo has argued that some Gymnopilus species have been observed to stain blue one growing season and not the next, suggesting that G. cyanopalmicola and G. palmicola may be the same species.

References

[1] Guzmán-Dávalos L, Mueller GM, Cifuentes J, Miller AN, Santerre A., Traditional infrageneric classification of Gymnopilus is not supported by ribosomal DNA sequence data, Mycologia, 2003. Nov-Dec;95(6):1204-14. PMID: 21149021

[2] https://mushroomexpert.com/gymnopilus_palmicola.html

[3] Guzmán-Dávalos L. (2006). "A New Bluing, Probably Hallucinogenic Species of Gymnopilus P. Karst. (Agaricomycetideae) from Mexico". International Journal of Medicinal Mushrooms. 8 (3): 289–293. doi:10.1615/intjmedmushr.v8.i3.110. ISSN 1521-9437.

[4] https://mushroomobserver.org/name/show_name/19773

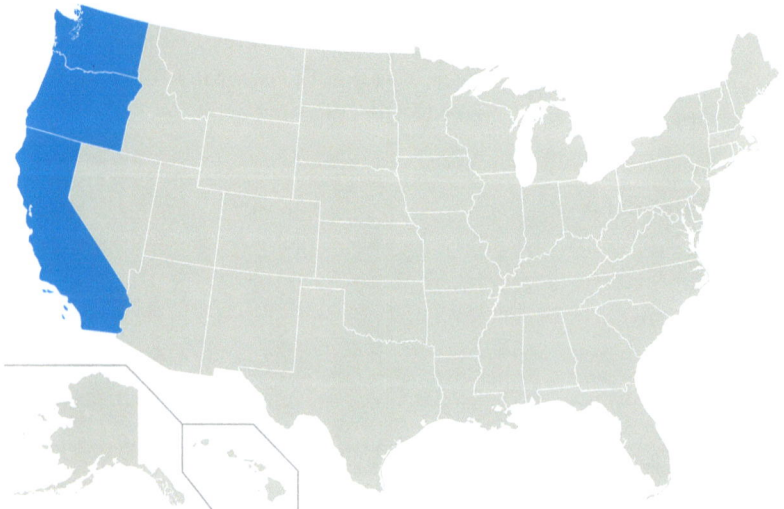

Cap Shape

Convex

Hymenium Shape

Gills

Hymenium Attachment

Adnate or Adnexed

Stipe Character

Ring

Psychoactive

?

Unknown

Ecological Type

Saprotrophic

Spore Print Color

Orange-Brown

The convex cap of Gymnopilus thiersii ranges from 3 to 15 cm in diameter and is purple-brown to reddish-brown in color. The adnexed or adnate gills are orange-brown. The stipe of this species ranges from 3 to 10 cm in length and 0.7 to 3.2 cm in width, with a similar purple-brown to reddish-brown colored cap flesh. Spore prints are orange-brown.

G. thiersii is a species of the large Gymnopilus genus that isn't described as thoroughly by mycologists as others in the genus. It has been known to grow on fallen pine logs in San Mateo County, California. The most recent observations show the mushrooms growing gregariously in small clusters in California.

Like many other Gymnopilus species, it stains blue-green when damaged, and thus likely contains psilocybin. G. thiersii has been described as nearly identical to closely-related species Gymnopilus luteofolius, but can purportedly be differentiated by its spore shape and subtle difference in coloration. Different species within the Gymnopilus genus are often difficult to differentiate by morphology alone, as many closely resemble one another. In addition, recent scientific evidence shows many physical features traditionally used to differentiate Gymnopilus species—like spore dimensions and the presence or absence of a partial veil—are not always genetically accurate. For the beginner in the field, it is more important to differentiate Gymnopilus species from the Galerina genus, which can be physically similar, but deadly if ingested. The Pholiota genus is also commonly confused with the Gymnopilus genus by morphology alone.

References

[1] Seidl MT, A New Species of Gymnopilus from Northern California, Mycotaxon, Volume 34, Pages 217-220, January 20, 1989.

[2] https://www.mushroomexpert.com/gymnopilus.html

[3] https://mushroomobserver.org/name/show_name/4943

[4] Guzmán-Dávalos L, Mueller GM, Cifuentes J, Miller AN, Santerre A., Traditional infrageneric classification of Gymnopilus is not supported by ribosomal DNA sequence data, Mycologia, 2003. Nov-Dec;95(6):1204-14. PMID: 21149021

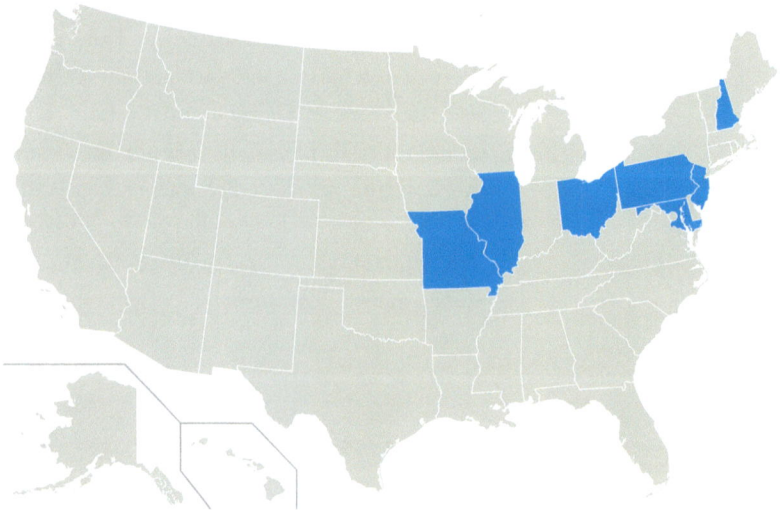

Cap Shape

Convex

Hymenium Shape

Gills

Hymenium Attachment

Adnate

Stipe Character

Cortina

Psychoactive

Low

Ecological Type

Saprotrophic

Spore Print Color

Orange-Brown

The broadly convex cap of the agaric mushroom species Gymnopilus validipes ranges from 4 to 30 cm in diameter. When the mushroom is young, the cap flesh is a tawny yellow, becoming rusty orange or reddish-orange as the mushroom ages. The hymenium attachment is adnate to slightly uncinate and thick. The gills are yellowish white and become browner as the mushroom matures. The stipe normally ranges from 100 to 250 mm in length and 25 to 50 mm in thickness, with a color ranging from yellowish white to rusty orange to yellow orange. There is a partial cortinate veil, and a ring on the stipe. Some reports have noted the stipe can reach a diameter of 5 cm, appearing swollen—differentiating it from other species within the same genus. G. validipes spore prints are orange-brown.

These fungi grow gregariously to cespitose on fallen hardwood, tree stumps, or wooden debris widely throughout North America, central Europe, and northern Europe in June through November.

Like other members of the genus Gymnopilus, G. validipes has been confirmed to contain psilocybin and has been described as weakly to mildly psychoactive. Unlike many other Gymnopilus species, which have a bitter taste, G. validipes has been described as having a mild taste. The first psychoactive reports from G. validipes came from people who had mistaken G. validipes for an edible honey mushroom, Armillaria mellea. This species of Gymnopilus belongs to the underwoodii-validipes phylogenetic clade, containing only three species: G. validipes, G. cf. flavidellus, and Gymnopilus underwoodii.

References

[1] Stamets, Paul (1996). Psilocybin Mushrooms of the World. Berkeley: Ten Speed Press. Pages 184 – 185. ISBN 0-9610798-0-0.

[2] https://mushroomobserver.org/name/show_name/2332

[3] https://mushroomobserver.org/name/show_name_description/5012

[4] https://www.shroomery.org/12474/Gymnopilus-validipes

[5] Hatfield GM, Valdes LJ, Smith AH, The Occurrence of Psilocybin in Gymnopilus species, Lloydia. 1978 Mar-Apr;41(2):140-4. PMID: 565861

[6] Guzmán-Dávalos L, Mueller GM, Cifuentes J, Miller AN, Santerre A., Traditional infrageneric classification of Gymnopilus is not supported by ribosomal DNA sequence data, Mycologia, 2003. Nov-Dec;95(6):1204-14. PMID: 21149021

GENUS PANAEOLUS

The genus Panaeolus describes a group of saprotrophic mushroom species that grow on dung, decaying forest debris, and decaying grass debris. There are roughly 30 described Panaeolus species worldwide. Panaeolus mushrooms are agaric, with fruiting bodies that allow for clearly differentiable caps and stipes by morphology. While it is possible for knowledgeable mushroom observers to differentiate Panaeolus species from mushrooms in other genera, it is difficult to differentiate individual Panaeolus species from one another. Parsing out species may therefore require the use of a microscope.

Panaeolus species are commonly found in Europe and North America, and most often grow on animal dung in fertile grasslands. A few Panaeolus species have been found growing in more tropical or warm weather climates like those of Hawaii, Cambodia, and Africa. These mushrooms are commonly referred to as "little brown mushrooms" due to their dull appearance. Common morphological characteristics of Panaeolus species are a gray, brown, or blackish cap that is conical, convex, or bell-shaped. In addition, Panaeolus have gills on the hymenium that sport a gray or black coloration as the mushrooms reach advanced age. The caps of Panaeolus species are relatively small compared to other mushrooms, and the spore print color is dark purple-brown or black. The stems of Panaeolus species are relatively long compared to their cap diameters in comparison to other types of mushrooms. Microscopically, Panaeolus spores have an apical germ pore. Even when macroscopic morphology and microscopic characteristics are taken into account, it is possible that these characteristics may result in mistaking the species for mushrooms within the genus Psathyrella. Some mycologists have developed a sulfuric acid-based test for differentiating Panaeolus from Psathyrella. Under a microscope, if the spores fade when mounted in sulfuric acid, the specimen is likely Psathyrella. If the spores do not fade under these conditions, the specimen is likely Panaeolus.

Mushrooms of the Panaeolus genus are psychoactive and contain the hallucinogenic compound psilocybin. Panaeolus species are part of the family Coprinaceae, alongside other genera Psathyrella and Coprinus. Most psilocybin-containing species within this family are in the genus Panaeolus. Notable Panaeolus species that have been found to contain psilocybin, psilocin, or both include: P. africanus, P. antillarum, P. cambodginiensis, P. castaneifolius, P. cyanescens, P.

fimicola, P. foenisecii, P. papilionaceus, P. semiovatus, P. subbalteatus, and P. tropicalis. Like other psilocybin-containing mushrooms, some of the hallucino-gen-containing Panaeolus undergo a bluing reaction, wherein the stipe or the cap flesh turn green or blue when handled. Historically and taxonomically, the species that bruise blue have been categorized into a separate subgroup called Copelandia. In this way, Copelandia and Panaeolus can often be considered synonyms at the genus level.

References
[1] Stamets, Paul (1996). Psilocybin Mushrooms of the World. Berkeley: Ten Speed Press. Page 67. ISBN 0-9610798-0-0.
[2] Maruyama T, Shirota O, Kawahara N, Yokoyama K, Makino Y, Goda Y, Discrimination of psychoactive fungi (commonly called "magic mushrooms") based on the DNA sequence of the internal transcribed spacer region, Shokuhin Eiseigaku Zasshi, 2003. Volume 44, Issue 1, pages 44-8.
[3] Gerhardt, E, Taxonomische Revision der Gattungen Panaeolus und Panaeolina (Fungi, Agaricales, Coprinaceae). Bibliotheca Botanica, 1996. Volume 147, pages 1-149. English translation from www.mushroomobserver.org
[4] Merlin MD, Allen JW, Species identification and chemical analysis of psychoactive fungi in the Hawaiian islands. Journal of Ethnopharmacology, 1993. Volume 40, Issue 1: Pages 21-40.
[5] Guzman G, Allen J, Gartz J, A Worldwide Geographical Distribution of the Neurotropic Fungi, an Analysis and Discussion, Annual Museo Civico di Rovereto, 1998. Volume 14, Pages 189-280.
[6] https://www.mushroomexpert.com/panaeolus.html
[7] https://mushroomobserver.org/species_list/show_species_list/1402

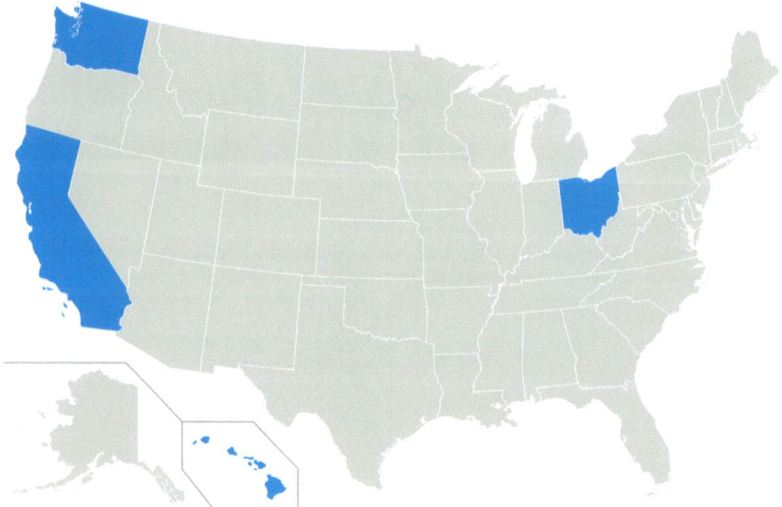

Cap Shape

Convex

Hymenium Shape

Gills

HymeniumAttachment

Adnexed

Stipe Character

Bare Stipe

Psychoactive

Unknown

Ecological Type

Saprotrophic

Spore Print Color

Black

The beige-toned, convex cap of Panaeolus bisporus, also known as Copelandia bisporus, ranges from 15 to 30 mm in diameter. The dark gray to black adnexed gills are crowded and contain a white edge. The rare mushroom's bare stipe ranges from 65 to 120 mm in length and ranges from 2 to 3 mm in width. The stipe color is a translucent gray outside, with a reddish-brown interior. The P. bisporus spore print color is black.

The mushroom grows gregariously on dung, and has been observed widely distributed throughout the world, in Hawaii, California, North Africa, Spain, and Switzerland.

Like other mushrooms in the Panaeolus genus, it bruises blue and contains psilocybin. The hallmark characteristics that P. bisporus shares with many other hallucinogenic Panaeolus species are growing on dung, having a hemispheric cap, and producing black spore prints. The Copelandia, which includes P. bisporus, as well as the species cambodginiensis, cholorocystis, cyanescens, and tropicalis, grow in tropical or semitropical climates and bruise bluish. P. bisporus strongly resembles these other species to the naked eye. Under a microscope, however, these species all have a singular characteristic pleurocystidia morphology, describing a large cell found on the gill face of the mushroom.

References

[1] https://mushroomobserver.org/name/show_name/24591

[2] Gerhardt, E, Taxonomische Revision der Gattungen Panaeolus und Panaeolina (Fungi, Agaricales, Coprinaceae). Bibliotheca Botanica, 1996. Volume 147, pages 1-149. English translation from www.mushroomobserver.org

[3] Stamets, Paul (1996). Psilocybin Mushrooms of the World. Berkeley: Ten Speed Press. Page 67. ISBN 0-9610798-0-0.

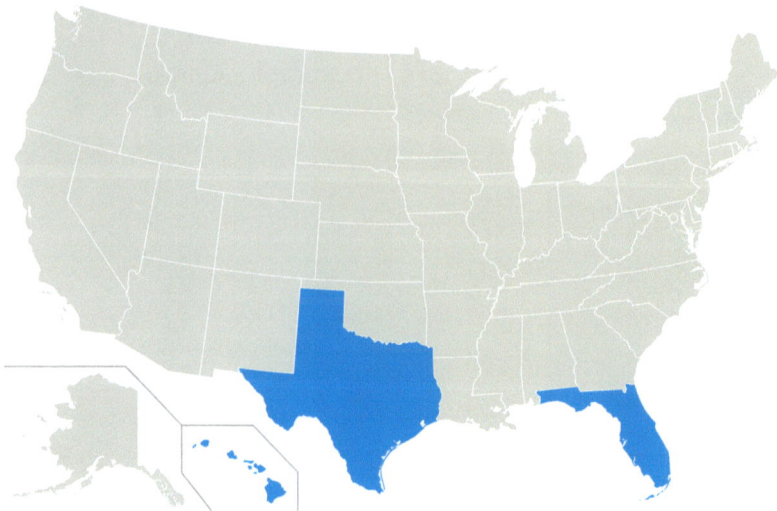

Cap Shape

Convex

Hymenium Shape

Gills

Hymenium Attachment

Adnexed

Stipe Character

Bare Stipe

Psychoactive

Medium

Ecological Type

Saprotrophic

Spore Print Color

Black

The brown, convex cap of Panaeolus cambodginiensis, also known as Copelandia cambodginiensis, ranges from 1.2 to 2.5 cm in diameter. The cap flesh is darker brown when the mushrooms are young, then fades to lighter brown with age. The adnexed gills of P. cambodginiensis are ascending, slightly uncinate, and grayish black to black. The bare stipe of this species ranges from 80 to 100 mm in length and 1.5 to 3 mm in thickness. The stipe color is whitish to cream in color, and, like other mushrooms in the Panaeolus genus, turns rapidly blue when damaged. The spore print color is black.

This species grows gregariously and can grow in scattered clusters. Like other Panaeolus species, it grows on dung, preferably buffalo and cattle dung. In the United States, P. cambodginiensis has been observed in Hawaii and Texas. It has also been described in Cambodia, Columbia, and the Asian subtropics. P. cambodginiensis is psychoactive and contains psilocybin.

Members of the Panaeolus genus are difficult to differentiate from one another by sight alone, as most present as little brown mushrooms. They are best differentiated at the cellular level of pleurocystidia with a microscope.

References

[1] https://www.shroomery.org/12483/Panaeolus-cambodginiensis

[2] https://www.mushroomexpert.com/panaeolus.html

[3] https://mushroomobserver.org/name/show_name_description/5142

[4] Merlin MD, Allen JW, Species identification and chemical analysis of psychoactive fungi in the Hawaiian islands. Journal of Ethnopharmacology, 1993. Volume 40, Issue 1: Pages 21-40.

[5] Stamets, Paul (1996). Psilocybin Mushrooms of the World. Berkeley: Ten Speed Press. Page 67. ISBN 0-9610798-0-0.

[6] Guzman G, Allen J, Gartz J, A Worldwide Geographical Distribution of the Neurotropic Fungi, an Analysis and Discussion, Annual Museo Civico di Rovereto, 1998. Volume 14, Pages 189-280.

PANAEOLUS (COPELANDIA) CYANESCENS

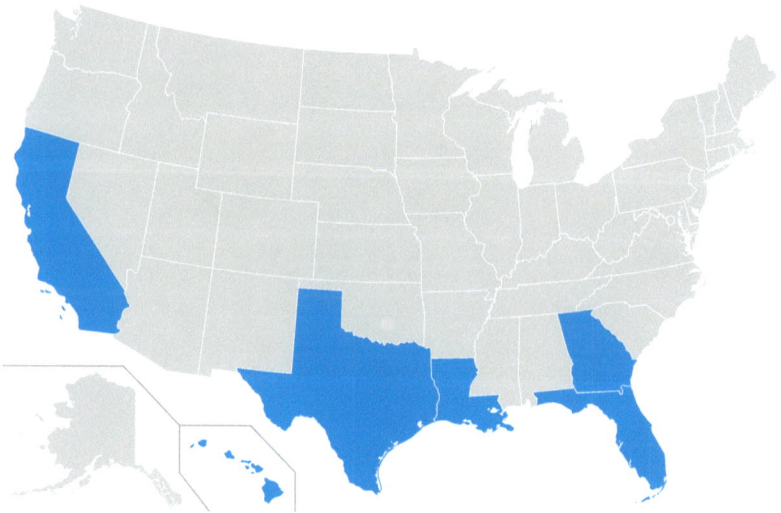

Cap Shape

Convex

Hymenium Shape

Gills

HymeniumAttachment

Adnate

Stipe Character

Bare Stipe

Psychoactive

Medium

Ecological Type

Saprotrophic

Spore Print Color

Black

The convex cap of Panaeolus cyanescens ranges from 1.5 to 4 cm in diameter and is light brown when the mushroom is young. With age, the cap becomes off-white to light gray in color. The broadly adnate gills are light gray when the mushroom is new and begin to turn black as the mushroom becomes older. The bare stipe grows to near 12 cm in length and ranges from 2 to 3 mm in thickness. Like the cap, the stem is light brown when the mushroom is young and becomes light gray to off-white with age.

P. cyanescens contains the psychoactive hallucinogen psilocybin, and its stem bruises blue when handled like some other psychoactive mushrooms. The spore print color of this species is black. P. cyanescens is coprophilous, meaning it grows on dung, and can grow in solitude, sparsely scattered, or gregariously in clusters. It is very widely distributed throughout tropical or near-tropical climates in the northern and southern hemispheres, especially Mexico, the Phillipines, Australia, Brazil, and Bolivia. In the United States, P. cyanescens is commonly found along the gulf coast in Louisiana and Florida, as well as Hawaii. The potency of this species has been noted as being moderate, and it has been found to contain psilocin as well as psilocybin.

References

[1] https://www.shroomery.org/12485/Panaeolus-cyanescens

[2] https://mushroomobserver.org/name/show_name/18361

[3] https://www.mushroomexpert.com/panaeolus.html

[4] Stamets, Paul (1996). Psilocybin Mushrooms of the World. Berkeley: Ten Speed Press. Page 75. ISBN 0-9610798-0-0.

[5] Maruyama T, Shirota O, Kawahara N, Yokoyama K, Makino Y, Goda Y, Discrimination of psychoactive fungi (commonly called "magic mushrooms") based on the DNA sequence of the internal transcribed spacer region, Shokuhin Eiseigaku Zasshi, 2003. Volume 44, Issue 1, pages 44-8.

PANAEOLUS CINCTULUS

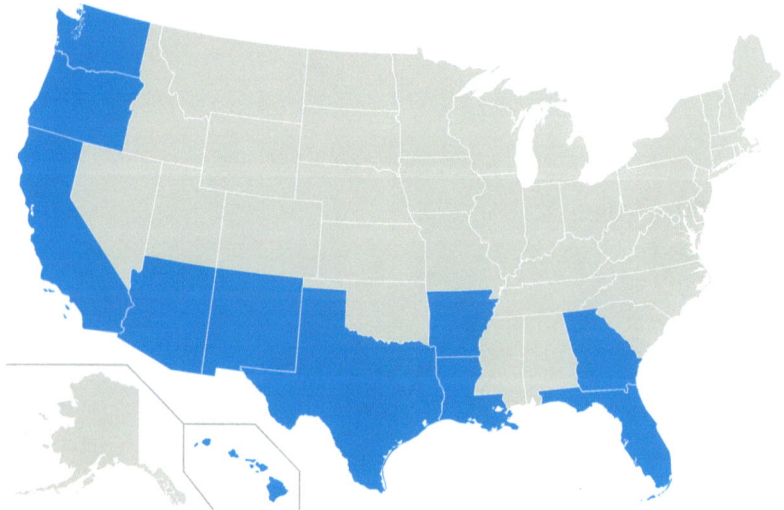

Cap Shape

Convex

Hymenium Shape

Gills

Hymenium Attachment

Adnate or Adnexed

Stipe Character

Bare Stipe

Psychoactive

Low

Ecological Type

Saprotrophic

Spore Print Color

Black

The cap of Panaeolus cinctulus is convex when the mushroom is young, then becomes broadly umbonate with age. The cap flesh is light, warm brown to cream-colored, and the cap ranges from 2 to 5 cm in diameter. The mottled brown gills have white edges and are adnate to uncinate and close together. The stipe ranges from 50 to 60 mm in length and 2 to 4 mm in thickness. The stem sometimes bruises in a blue hue at the base, or when handled. The stipe is reddish brown to white in color and is hollow and bare. The P. cinctulus spore print is jet black.

The mushrooms grow gregariously and are ubiquitous in grassy areas, especially those that have been fertilized. P. cinctulus grows well on dung, particularly in areas where horses frequent, and on hay where bales are left out to rot. This species is widely distributed throughout North America, South America, Europe, middle Siberia, and Hawaii.

Mushroom experts have noted this may be the world's most widely-distributed psychoactive, psilocybin-containing mushroom.

References

[1] https://mushroomobserver.org/name/show_name_description/1266
[2] https://www.shroomery.org/12484/Panaeolus-cinctulus

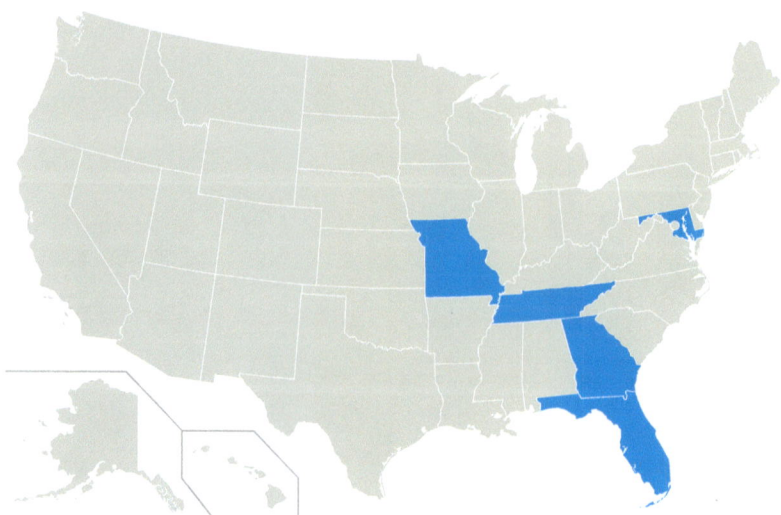

Cap Shape

Convex

Hymenium Shape

Gills

Hymenium Attachment

Adnate

Stipe Character

Bare Stipe

Psychoactive

Low

Ecological Type

Saprotrophic

Spore Print Color

Black

The cap shape of Panaeolus fimicola is convex and may expand to flat with age. The cap color is dingy gray to black with a diameter ranging from 1.5 to 4 cm. The mushroom's adnate gills are closer together and crowded and have a gray color when the mushroom is new. With age, the gills become blacker, but retain white edges. The stipe is bare, and ranges from 4 to 10 cm in length and 1 to 3 mm in thickness. The stipe is a dingy white to clay-colored hue. Like many other mushrooms in the Panaeolus genus, P. fimicola spore prints are black.

The mushrooms grow in solitude or in scattered groups in fertilized soil, dung, or grassy places. This species is common throughout North and South America, Africa, and Europe, but has been noted to be rarely-described due to its very close resemblance to other Panaeolus species. Recent observations have found these mushrooms in Georgia, Missouri, and Florida within the United States.

Panaeolus species are difficult to differentiate from one another without a microscope, and are commonly called "little brown mushrooms." This species has also been mistaken for mushrooms within the Psilocybe genus. The mushroom is a latent producer of hallucinogen psilocybin and has been noted to be weakly psychoactive with regard to potency. Due to this latent production, psilocybin has been detected in some collections and not others.

References

[1] https://www.shroomery.org/12486/Panaeolus-fimicola
[2] https://www.mushroomexpert.com/panaeolus.html
[3] https://mushroomobserver.org/name/show_name/1141
[4] Stamets, Paul (1996). Psilocybin Mushrooms of the World. Berkeley: Ten Speed Press. Page 76. ISBN 0-9610798-0-0.

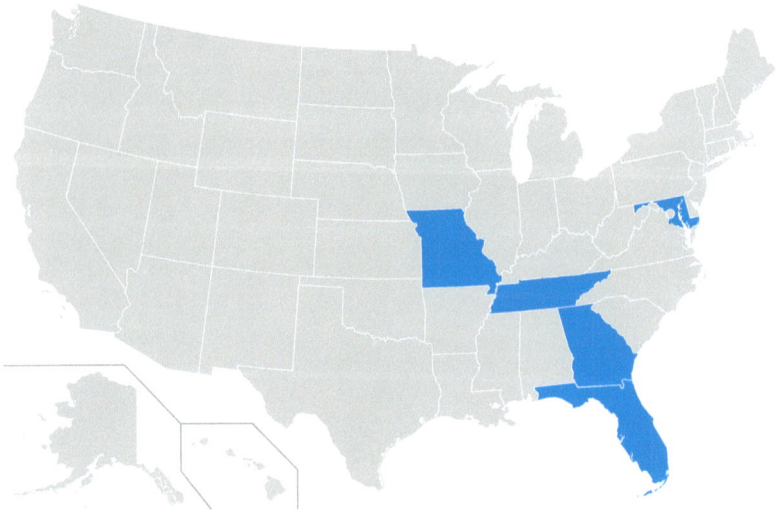

Cap Shape

Convex or Campanulate

Hymenium Shape

Gills

Hymenium Attachment

Adnate or Adnexed

Stipe Character

Bare Stipe

Psychoactive

Unkown

Ecological Type

Saprotrophic

Spore Print Color

Black or Olive Green

The cap diameter of Panaeolus olivaceus ranges from 1 to 4 cm. The cap shape can exhibit both a flat, convex, conical, or campanulate shape within the same P. olivaceus species. Olive colored, black, brown, and grayish brown caps have been observed. The adnexed or adnate gills range from gray-brown to olive-black in color. The mushroom's bare stipe ranges from 3 to 8 cm in length and 1.5 to 3 mm in thickness. Stem color is light beige to grayish brown and is entirely covered with white flakes when the mushroom is young. Like other Panaeolus species in the genus, spore prints have been described as both black or olive green. The presence of lemon-shaped, flattened, finely rough spore prints is a primary diagnostic characteristic of this species when viewed under a microscope. Panaeolus fimicola appears similar, but exhibits smooth spores in a spore print.

P. olivaceus can grow gregariously or sparsely scattered and has been observed in areas rich in fertilized grass, horse dung, and bird droppings. It is distributed throughout North and South America. Within the United States, it has been collected in Washington, Oregon, Florida, and Georgia. The mushroom is psychoactive but specific data on potency has not been published.

References

[1] https://mushroomobserver.org/observer/observation_search?pattern=Panaeolus+olivaceus

[2] https://mushroomobserver.org/name/show_name/18360

[3] https://mushroomobserver.org/name/show_name_description/5131

[4] https://www.mushroomexpert.com/panaeolus.html

GENUS PHOLIOTINA

Pholiotina describes a genus comprised of over 62 species of small, agaric mushrooms that share taxonomic characteristics. As the Pholiotina genus is polyphyletic, meaning many of its species are grouped together but do not share a common evolutionary ancestor, it is likely that some of the current species considered Pholiotina will be categorized into different genera in the future. Historically, some mushroom researchers considered the Pholiotina genus to be part of the larger Conocybe genus. In contradiction with this school of thought, a phylogenetic analysis showed the Pholiotina genus to truly comprise a molecularly distinct set of organisms. In spite of this study, certain Pholiotina species, like Pholiotina cyanopus and Pholiotina smithii, are sometimes referred to as Conocybe cyanopus and Conocybe smithii. Another notable common Pholiotina species, Pholiotina rugosa, is called Conocybe filaris in most common mushroom observation field guides. Originally, Pholiotina species were differentiated from Conocybe species by the phenotypic presence of partial veils on the Pholiotina specimens. More recently, the Pholiotina genus has been expanded to include species lacking partial veils.

Pholiotina species generally have small caps compared to other similar mushroom genera. Many mushrooms in the genus have dry caps, rusty brown or orange spore print colors, and thick-walled, smooth spores and contain a germ pore, or small pore in the spore wall where the germ tube protrudes during germination. Two prominent species of Pholiotina are hallucinogenic and contain the compound psilocybin. Morphologically, these Pholiotina species, P. cyanopus and P. smithii, have conical to convex caps, adnate gills, and brown spore prints. These mushrooms are saprotrophic and grow mostly within temperate climates in North America and Europe in moist areas like riverbanks, lawns, bogs, grasses, and patches of moss. The related species mentioned above, P. rugosa, grows throughout the Pacific Northwest on wood chips, soil, and compost. This species is not psychoactive and is instead very poisonous and deadly if consumed. It is difficult to differentiate the psychoactive Pholiotina species from one another macroscopically, but it is possible to do so reliably under a microscope on the basis of spore size. It is important for mushroom observers to be able to differentiate poisonous Galerina species and closely-related poisonous P. rugosa from P. cyanopus and P. smithii, which can only truly reliably be done under a microscope by mushroom experts.

References

[1] Halama M, Poliwoda A, Jasicka-Misiak I, Wieczorek P, Rutkowski R, Pholiotina cyanopus, a rare fungus producing psychoactive tryptamines, Open Life Sciences, 2014. Volume 10 Issue 1, pages 40-51.

[2] Zhuk O, Jasicka-Misiak I, Poliwoda A, Kazakova A, Godovan VV, Halama M, Wieczorek PP, Research on acute toxicity and the behavioral effects of methanolic extract from psilocybin mushrooms and psilocin in mice, Toxins, 2015. Volume 7 Issue 4, pages 1018-29. 7(4):1018-29

[3] Stamets, Paul (1996). Psilocybin Mushrooms of the World. Berkeley: Ten Speed Press. Page 177. ISBN 0-9610798-0-0.

[4] https://www.inaturalist.org/taxa/118269-Pholiotina

[5] https://www.zobodat.at/pdf/OestZPilz_16_0133-0145.pdf

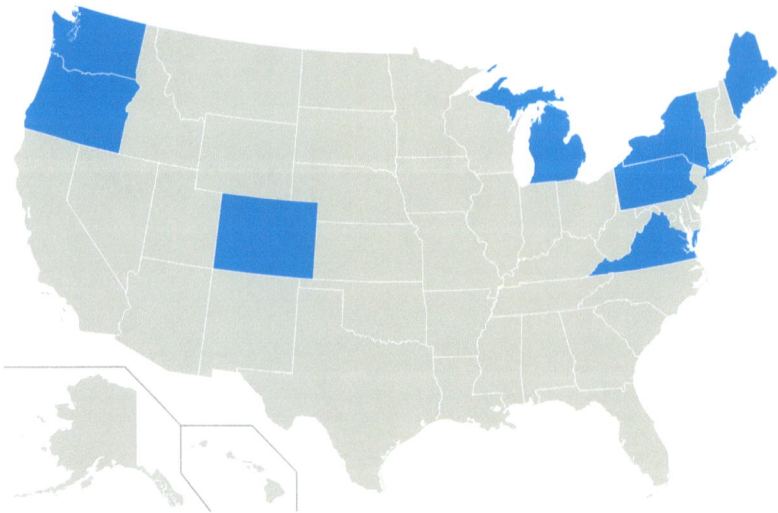

Cap Shape

Convex

Hymenium Shape

Gills

Hymenium Attachment

Adnate

Stipe Character

Cortina

Psychoactive

High

Ecological Type

Saprotrophic

Spore Print Color

Brown

The cap of Pholiotina cyanopus has a convex or conical shape that becomes broadly convex with age. The cap ranges from 0.7 cm to 2.5 cm in diameter and has a cinnamon brown to burnt orange color. The mushroom's adnate gills are close and moderately broad, with a cinnamon brown color and white edges. The gills darken as the mushroom ages, becoming a rusty brown hue. The stipe ranges from 20 to 40 mm in length and 1 to 1.4 mm in thickness. The stipe has a partial veil that is thinly cortinate, and a whitish color. Like other psilocybin-containing mushrooms, the stipe bruises blue or blue-green. The spore print color of P. cyanopus is rusty brown.

The mushrooms grow in solitude or gregariously from early summer until early fall and are distributed throughout temperate climates in North America and Europe. This mushroom has been reported to be rare in Europe. Within the United States, P. cyanopus has been observed in Colorado, Michigan, New York, Oregon, and Washington.

The species is psychoactive, containing psilocybin, psilocin, baeocystin, norbaeocystin, and aeruginascin. The concentration of psilocybin has been published as being relatively high, 0.90% by dry weight.

References

[1] https://mushroomobserver.org/name/show_name/20054

[2] https://www.shroomery.org/12461/Pholiotina-cyanopus

[3] Halama M, Poliwoda A, Jasicka-Misiak I, Wieczorek P, Rutkowski R, Pholiotina cyanopus, a rare fungus producing psychoactive tryptamines, Open Life Sciences, 2014. Volume 10 Issue 1, pages 40-51.

[4] Zhuk O, Jasicka-Misiak I, Poliwoda A, Kazakova A, Godovan VV, Halama M, Wieczorek PP, Research on acute toxicity and the behavioral effects of methanolic extract from psilocybin mushrooms and psilocin in mice, Toxins, 2015. Volume 7 Issue 4, pages 1018-29. 7(4):1018-29

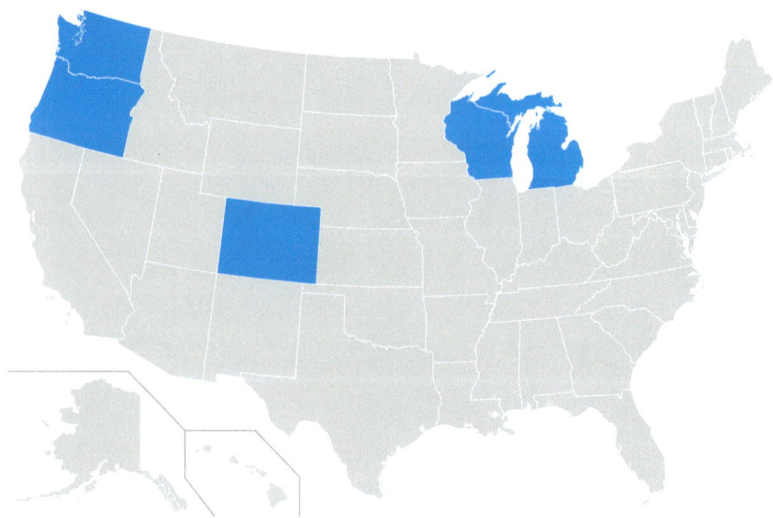

Cap Shape

Conical

Hymenium Shape

Gills

Hymenium Attachment

Adnate

Stipe Character

Bare Stipe

Psychoactive

High

Ecological Type

Saprotrophic

Spore Print Color

Brown

The cap of Pholiotina smithii can appear conical to convex, and becomes nearly planar as the mushroom reaches an advanced age. The cap ranges from 0.3 to 1.3 cm in diameter and has a cinnamon brown color. The adnate to adnexed gills of the mushroom are crowded and colored a pale, grayish yellow. As the mushroom ages, the gills darken to a rusty brown color. The bare stipe ranges from 1 to 7.5 cm in length and 0.75 to 1.5 mm in thickness. The stem's color is a pale white, which becomes blue with bruising or age. Like other hallucinogenic mushrooms, P. smithii bruises blue when handled. The spore print color of this species is brown.

P. smithii grows singularly or in scattered groups or small clusters in bogs, ditches, swampy areas, clumps of moss, river banks, and lawns. The mushroom has been observed growing in temperate climates throughout North America, Central Europe, and Northern Europe. Within the United States, it has been observed in Washington, Oregon, and Colorado.

P. smithii is psychoactive and contains the hallucinogens psilocybin, psilocin, and baeocystin. Observers have noted that the species is potently psychoactive. Some older mycology texts refer to P. smithii and related species Pholiotina cyanopus by the genus name Conocybe. The cyanopus species is closely related and morphologically similar, but can be differentiated from P. smithii because the gills of P. cyanopus lack the cinnamon flush color of P. smithii. The species can also be differentiated by spore size under a microscope.

References

[1] https://mushroomobserver.org/name/show_name_description/5124

[2] https://www.shroomery.org/12489/Pholiotina-smithii

[3] Halama M, Poliwoda A, Jasicka-Misiak I, Wieczorek P, Rutkowski R, Pholiotina cyanopus, a rare fungus producing psychoactive tryptamines, Open Life Sciences, 2014. Volume 10 Issue 1, pages 40-51.

[4] Zhuk O, Jasicka-Misiak I, Poliwoda A, Kazakova A, Godovan VV, Halama M, Wieczorek PP, Research on acute toxicity and the behavioral effects of methanolic extract from psilocybin mushrooms and psilocin in mice, Toxins, 2015. Volume 7 Issue 4, pages 1018-29. 7(4):1018-29

[5] Stamets, Paul (1996). Psilocybin Mushrooms of the World. Berkeley: Ten Speed Press. Page 177. ISBN 0-9610798-0-0.

GENUS PLUTEUS

The relatively vast Pluteus genus of small, agaric mushrooms contains over 300 distinct species. Studied mushroom observers can use conserved general macroscopic and microscopic characteristics of Pluteus species to differentiate them from other genera. In general, Pluteus species are saprotrophic and grow on decaying wood or wood chips. The spore print of this species is pink or brownish pink, and the gills of this species grow freely apart from the stipe. Mushroom observer Paul Stamets described Pluteus generally as having a planar to convex cap, free, pink gills, and a ringless stipe that freely breaks away from the cap when handled. It is possible for a knowledgeable mushroom observer to identify Pluteus species from other genera without a microscope based on the color and texture of the cap and stem, the size and shape of the cap, and the environment where the mushrooms are found growing. Differentiating Pluteus species from one another, however, is difficult macroscopically. For identification between Pluteus species, a sample should be mounted on a microscope slide in potassium hydroxide and cellular features like the pileipellis, pleurocystidia, cheilocystidia, and spores should be observed and characterized.

The most well known Pluteus species is Pluteus cervinus, commonly called the deer shield or fawn mushroom. This common species grows on rotten logs, is edible but not often used as a food source, and contains within the species several morphologically differing subspecies. The Pluteus genus contains a few psychoactive, psilocybin-containing species, which undergo the bluing reaction common to hallucinogenic mushrooms. Some mushroom observers have noted that Pluteus species do not exhibit the same traditional blueish hue as other genera of psilocybin-containing mushrooms, describing instead a dull blue, olive, blue-gray, green-brown, and blackish-green reaction. Some notable psilocybin-containing species are Pluteus salicinus, Pluteus villosus, Pluteus cyanopus, Pluteus americanus, and Pluteus glaucus. Pluteus species have been observed all over the world, from Brazil to Russia, to Europe and North America.

References

[1] Stamets, Paul (1996). Psilocybin Mushrooms of the World. Berkeley: Ten Speed Press. Page 190-191. ISBN 0-9610798-0-0.

[2] https://www.mushroomexpert.com/pluteus_americanus.html

[3] http://mushroomexpert.com/pluteus.html

[4] Justo A, Malysheva E, Bulyonkova T, Vellinga E, Cobian G, Nguyen N, Minnis A, Hibbett D, Molecular phylogeny and phylo-geography of Holarctic species of Pluteus section Pluteus (Agaricales: Pluteaceae), with description of twelve new species, Phytotaxa, 2014. Pages 1-85.

[5] https://mushroomobserver.org/species_list/show_species_list/1341

PLUTEUS AMERICANUS

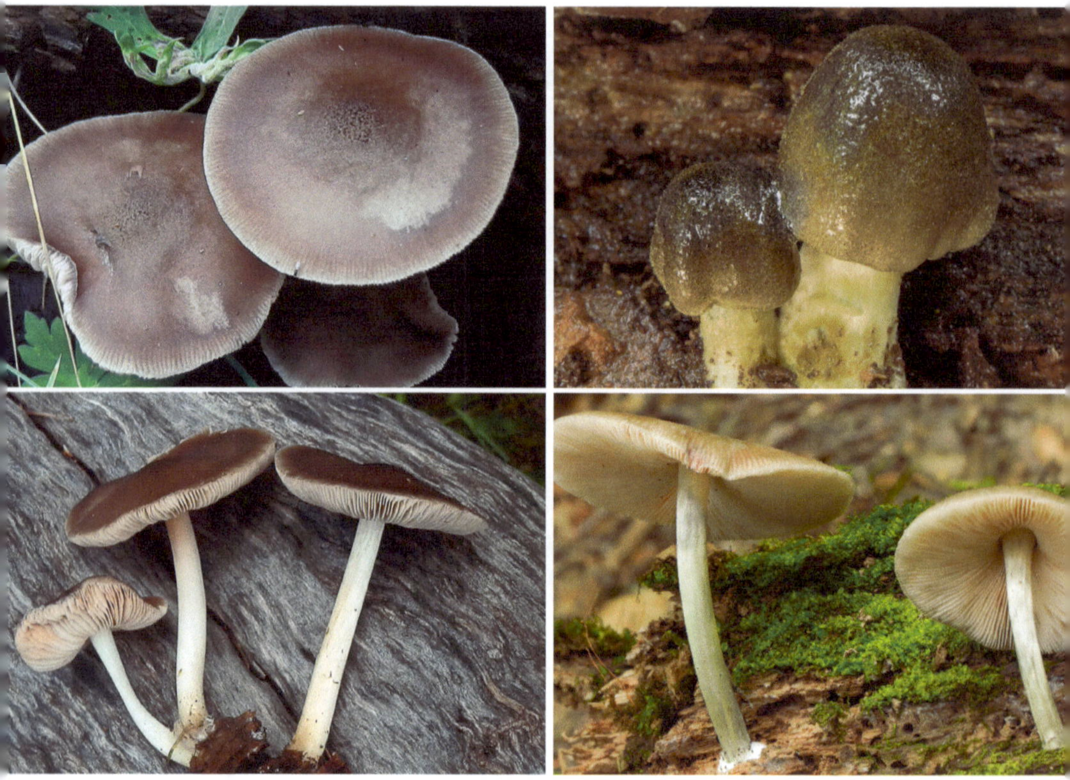

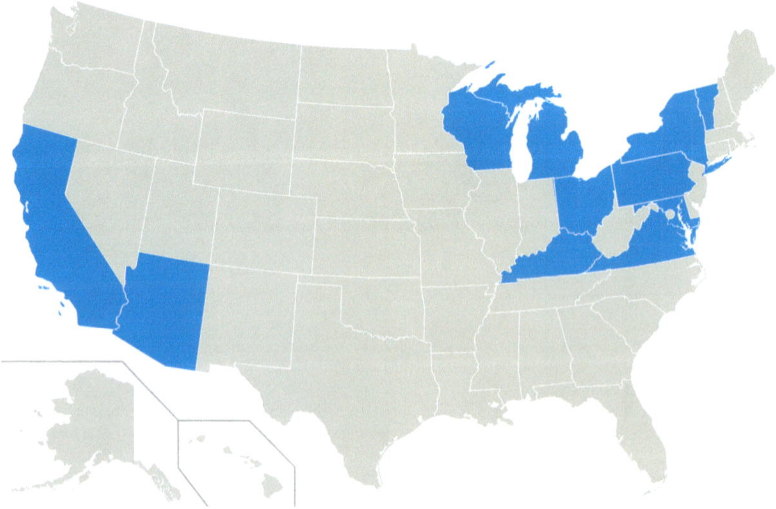

Cap Shape

Convex

Hymenium Shape

Gills

Hymenium Attachment

Free

Stipe Character

Bare Stipe

Psychoactive

Low

Ecological Type

Saprotrophic

Spore Print Color

Pink

The broadly convex cap of Pluteus americanus ranges from 4 to 7 cm in diameter and is a dark gray-brown when the mushroom is young. As the mushroom ages or dries, the cap can expand to plano-convex and the color turns pale gray with a dark center. The mushroom's free gills are whitish pink when the mushroom is young, turning brownish pink and bruising blue when handled. The bare stipe ranges from 4 to 6 cm in length and 3 to 5 mm in thickness and is white when the mushroom is young. As the mushroom ages, the stipe color becomes brownish and bruises gray-blue when handled. The P. americanus spore print is pink or brownish pink.

The species grows in solitude or gregariously throughout the United States east of the Rocky Mountains, as well as Eastern Russia. The mushrooms grow on the dead wood of hardwoods in hardwood-dominated or mixed forests in temperate climates.

The species is psychoactive. P. americanus is one of the few species of the Pluteus genus that bruises blue on the stem and gills when the mushroom is fresh. P. americanus was previously called Pluteus salicinus before it was determined that P. salicinus growth is restricted to Eurasia.

References

[1] Stamets, Paul (1996). Psilocybin Mushrooms of the World. Berkeley: Ten Speed Press. Page 190-191. ISBN 0-9610798-0-0.

[2] https://www.mushroomexpert.com/pluteus_americanus.html

[3] http://mushroomexpert.com/pluteus.html

[4] https://mushroomobserver.org/name/show_name/44676

[5] Justo A, Malysheva E, Bulyonkova T, Vellinga E, Cobian G, Nguyen N, Minnis A, Hibbett D, Molecular phylogeny and phylogeography of Holarctic species of Pluteus section Pluteus (Agaricales: Pluteaceae), with description of twelve new species, Phytotaxa, 2014. Pages 1-85.

PLUTEUS PHAEOCYANOPUS

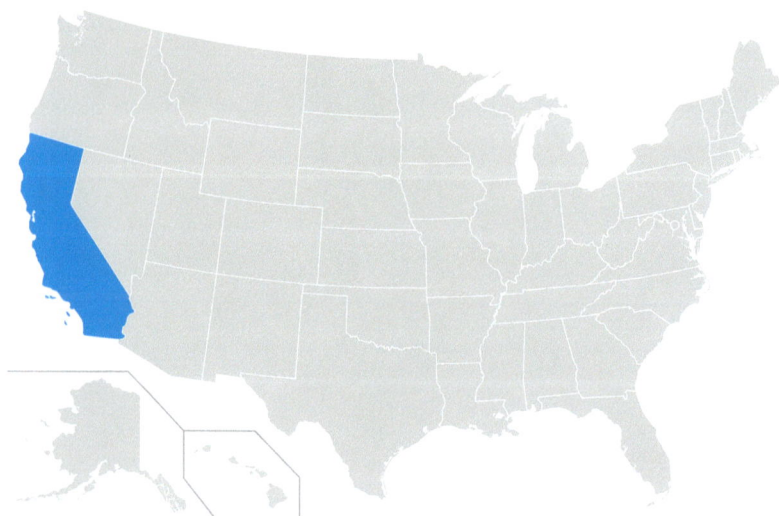

Cap Shape

Convex

Hymenium Shape

Gills

Hymenium Attachment

Free

Stipe Character

Bare Stipe

Psychoactive

Unknown

Ecological Type

Saprotrophic

Spore Print Color

Translucent or Pale Yellow

The convex cap of Pluteus phaeocyanopus ranges from 2 to 4 cm in diameter, and can appear plano-convex to planar. The cap is a brown color, which can range from dark watery brown to buckthorn brown with a darker color in the center. The free gills of P. phaeocyanopus are pale and whitish. The stipe ranges from 2 to 5 cm in length and 4 to 8 mm in width and has a gray green or gray olive color. The mushroom contains psilocybin. The spore print of this species has been observed in the main scientific publications to be translucent, pale yellow in potassium hydroxide (KOH).

In the United States, the mushrooms have only been documented to grow in California. They grow in solitude or gregariously on the dead wood of trees within the Quercus genus. P. phaeocyanopus was originally described by mycologist Harry Thiers as being another closely-related species, Pluteus cyanopus. Microscope examination by Minnis et al. revealed that the P. phaeocyanopus species warranted its own taxonomic species designation. Microscopically, P. phaeocyanopus can be differentiated from P. cyanopus by pleurocystidial shape and the presence of pigmentation in the lamellar cystidia of the two species.

References

[1] Stamets, Paul (1996). Psilocybin Mushrooms of the World. Berkeley: Ten Speed Press. Page 190-191. ISBN 0-9610798-0-0.

[2] Justo A, Malysheva E, Bulyonkova T, Vellinga E, Cobian G, Nguyen N, Minnis A, Hibbett D, Molecular phylogeny and phylogeography of Holarctic species of Pluteus section Pluteus (Agaricales: Pluteaceae), with description of twelve new species, Phytotaxa, 2014. Pages 1-85.

[3] https://mushroomobserver.org/observer/observation_search?pattern=Pluteus+phaeocyanopus

[4] Minnis A, Sundberg W, Pluteus section Celluloderma in the U.S.A., North American Fungi, 2010. Volume 5, Issue 1, Page 44.

GENUS PSILOCYBE

The genus Psilocybe contains an estimated 100 too 200 species of agaric gilled mushrooms that grow all over the world. Many Psilocybe species contain hallucinogenic compounds, resulting in this genus representing one of the most well-studied genera commonly referred to as "magic mushrooms." Psilocybe species are similar in macroscopic morphology to the other genera Hypholoma and Stropharia, and it is difficult to differentiate species within these genera from one another without the use of microscopy. These genera are all part of the Strophariaceae family, which also contains the psychoactive species-containing genus Pholiota. Recent taxonomic studies have shown that the Psilocybe genus is polyphyletic, meaning there are two distinct clades within the wider genus that are not directly related to one another. Like other mushroom genera, the classifications within the Psilocybe genus will likely be updated and shift in the coming years as sequencing technology and phylogenetic classification techniques improve.

Psilocybe mushrooms grow on dead or decaying matter, making them saphrotropic. They grow in a vast swath of different habitats, gravitating toward moisture. Psilocybe species grow on animal dung, decaying wood chips or dead trees, as well as mosses and soils. This genus is located in tropical climates and temperate climates in both wilderness and landscaped regions. The greatest diversity of species is found in South America, particularly in Chile and Brazil. In general, many species have brown conical to convex caps, adnate, adnexed or subdecurrent brown gills with white edges, and purple spore print coloration. Microscopically, Psilocybe species can be differentiated by their distinct apical germ pore on mounted spores, the absence of chrysocystidia, and a smooth elliptical spore shape. Like other psychoactive, psilocybin-containing species of mushroom, hallucinogenic Psilocybe species undergo a bluing reaction when handled. The most commonly recognized psychoactive species of Psilocybe are the hallucinogenic species Psilocybe semilanceata (commonly known as the "liberty cap") and Psilocybe cubensis. Other notable psychoactive species include Psilocybe cyanescens, Psilocybe mexicana, and Psilocybe azurescens. Commercially and historically, P. cubensis spores have been sold by head shops in suspension as microscopic specimens for observation, when they were actually most commonly used for cultivation.

References

[1] Borovička J, Oborník M, Stříbrný J, Noordeloos ME, Parra LA, Sánchez, Gryndler M, Phylogenetic and chemical studies in the potential psychotropic species complex of Psilocybe atrobrunnea with taxonomic and nomenclatural notes, Persoonia, 2015. Volume 34, Pages 1-9.

[2] Stamets, Paul (1996). Psilocybin Mushrooms of the World. Berkeley: Ten Speed Press. Pages 84-85. ISBN 0-9610798-0-0.

[3] https://www.mushroomexpert.com/stropharioid.html

[4] https://mushroomobserver.org/name/show_name_description/635

[5] https://www.sciencedirect.com/topics/neuroscience/psilocybe

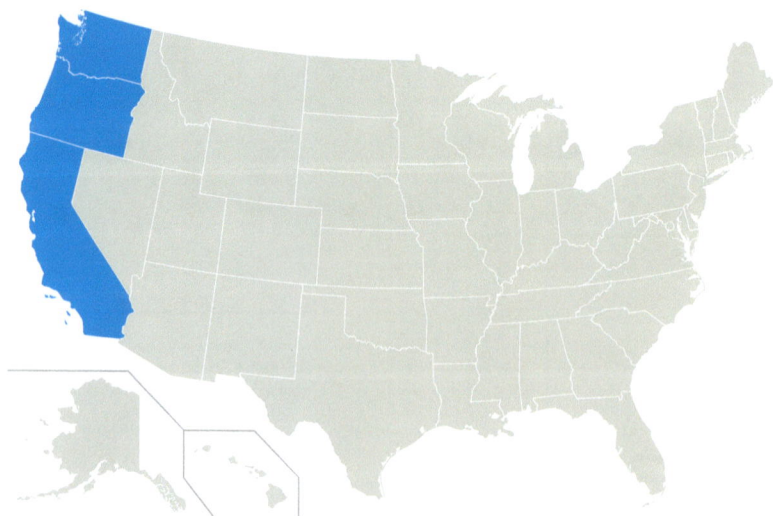

Cap Shape

Convex

Hymenium Shape

Gills

Hymenium Attachment

Adnate

Stipe Character

Ring

Psychoactive

Medium

Ecological Type

Saprotrophic

Spore Print Color

Brown

Psilocybe allenii has a convex cap that ranges from 2 to 4.5 cm in diameter and has a brown or yellow-brown color. The mushroom's adnate gills are a pale cinnamon brown color and become darker grayish brown with age as the lamellae become mottled with spores. The mushroom has a partial fibrillose veil. This species has a stipe that ranges from 3 to 6 cm in length and 3 to 6 mm in thickness, with a white surface that bruises blue. The spore print color is noted in the published literature as brown, and has been described as purple-brown to purple-gray and purple-black.

P. allenii grows gregariously or in small clusters on woody debris like wood chips and hardwood mulches in cold, fall weather between September and January. The species grows throughout the northwest region of North America. Within the United States, the species has been observed growing in the region spanning from California to Canada near British Columbia.

The mushroom is psychoactive and contains the hallucinogen psilocybin. P. allenii is morphologically very similar to Psilocybe cyanescens, as well as molecularly similar under a microscope. The P. allenii species, however, can be differentiated by its convex to hemispheric pileus and has been confirmed by DNA sequencing to be a distinct species.

References

[1] Stamets P, Mycelium Running: How Mushrooms Can Help Save the World, Ten Speed Press, 2005. Page 286.

[2] Borovička J, Oborník M, Stříbrný J, Noordeloos ME, Parra LA, Gryndler M, Phylogenetic and chemical studies in the potential psychotropic species complex of Psilocybe atrobrunnea with taxonomic and nomenclatural notes, Persoonia, 2015. Volume 34, Pages 1-9.

[3] https://www.shroomery.org/12492/Psilocybe-allenii

[4] Borovička J, Rockefeller A, Werner P, Psilocybe allenii - A new bluing species from the Pacific Coast, USA, Czech Mycology, 2012. Volume 64, Issue 2, Pages 181-195.

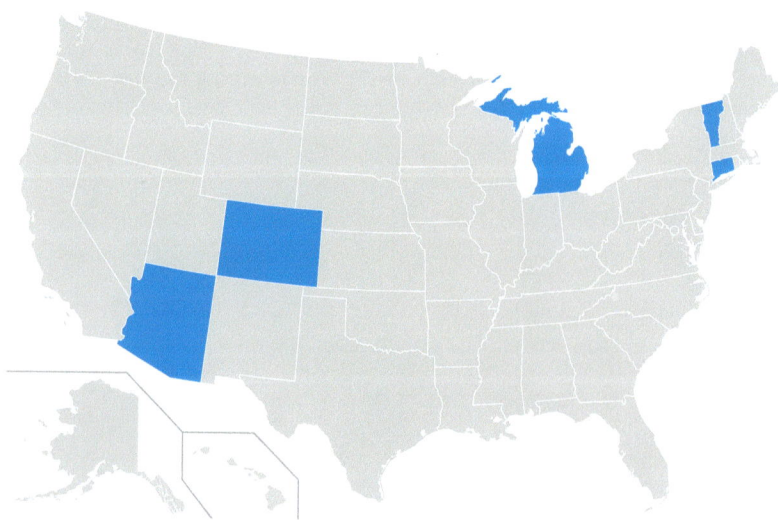

Cap Shape

Convex or Campanulate

Hymenium Shape

Gills

Hymenium Attachment

Adnate or Adnexed

Stipe Character

Cortina

Psychoactive

High

Ecological Type

Saprotrophic

Spore Print Color

Purple

The convex cap of Psilocybe aztecorum ranges from 1.5 to 3.5 cm in diameter. The cap appears obtusely conic or campanulate when the mushroom is young, maturing to be broadly convex or planar with advanced age. The mushroom's cap is colored dark chestnut brown and becomes yellow or off-white when the mushroom is dry. The cap bruises blue, especially around the margin. The adnate to adnexed gills are a light purple-gray color that can range to dark purple-brown. The stipe ranges from 55 to 75 mm in length and 3 to 4 mm in thickness and is colored white or gray. Like the cap, the stem easily bruises with a blueish color. There is a partial cortinate veil which can sometimes leave an annular zone in the upper stipe. The P. aztecorum spore print is purple.

This species grows gregariously or in small clusters in soil containing wood debris. The species has been observed to grow in groups of 5 to 20 mushrooms that sometimes occur in bundles. Occasionally, it can grow on pinecones and open woods with rich grasses. The species contains the hallucinogen psilocybin, and has been noted to be a fairly potent species.

References

[1] https://www.shroomery.org/12494/Psilocybe-aztecorum

[2] Borovička J, Oborník M, Stříbrný J, Noordeloos ME, Parra LA, Gryndler M, Phylogenetic and chemical studies in the potential psychotropic species complex of Psilocybe atrobrunnea with taxonomic and nomenclatural notes, Persoonia, 2015. Volume 34, Pages 1-9.

[3] Stamets, Paul (1996). Psilocybin Mushrooms of the World. Berkeley: Ten Speed Press. Pages 92-93. ISBN 0-9610798-0-0.

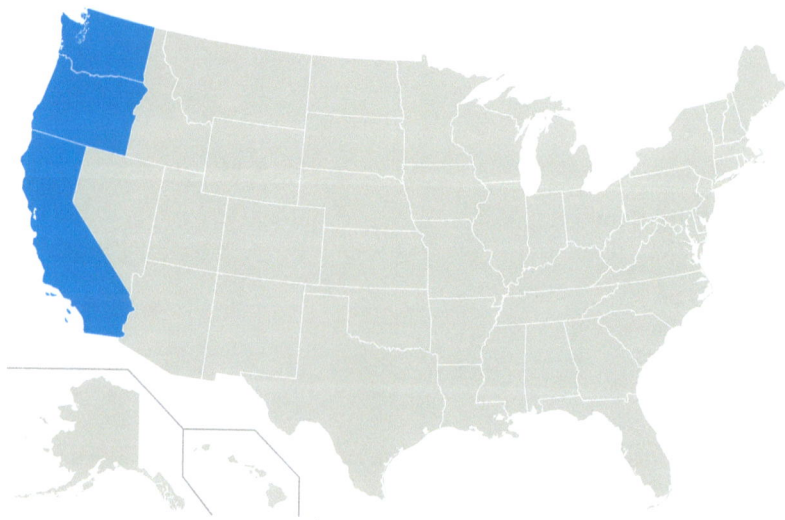

Cap Shape

Conical to convex

Hymenium Shape

Gills

Hymenium Attachment

Adnate or Sinuate

Stipe Character

Cortina

Psychoactive

High

Ecological Type

Saprotrophic

Spore Print Color

Purple

The cap of mushroom species Psilocybe azurescens ranges from 3 to 10 cm in diameter and can appear conical to convex. The cap expands to broadly convex as the mushroom ages and has a brown, caramel coloration. The cap has been observed to sometimes become pitted with blue or black zones of bruising. The brown gills can appear sinuate or adnate and, like the cap, stain indigo black when they are bruised. The mushroom's stipe ranges from 90 to 200 mm in length and 3 to 6 mm in thickness. The stem's color can range from silky white to dingy brown at the base. The spore print of this species is a dark purple, or purple black color.

P. azurescens grows in dense clusters and gregariously on deciduous wood chips or in soils rich in decaying wood debris and near dune grasses. In the United States, it is distributed along the north coast of Oregon near the shoreline, as well as in Washington.

The mushroom is psychoactive and contains the hallucinogen psilocybin. Researchers have noted this species is one of the most potent of the tryptamine-bearing mushrooms, with up to 1.8% psilocybin. Where P. azurescens grows in a dense mat, it causes wood to whiten. This species is one of the most potent in the world and exhibits one of the strongest bluing reactions.

References

[1] Borovička J, Oborník M, Stříbrný J, Noordeloos ME, Parra LA, Gryndler M, Phylogenetic and chemical studies in the potential psychotropic species complex of Psilocybe atrobrunnea with taxonomic and nomenclatural notes, Persoonia, 2015. Volume 34, Pages 1-9.

[2] Stamets, Paul (1996). Psilocybin Mushrooms of the World. Berkeley: Ten Speed Press. Pages 94-96. ISBN 0-9610798-0-0.

[3] https://www.shroomery.org/12495/Psilocybe-azurescens

[4] https://mushroomobserver.org/name/show_name/5346

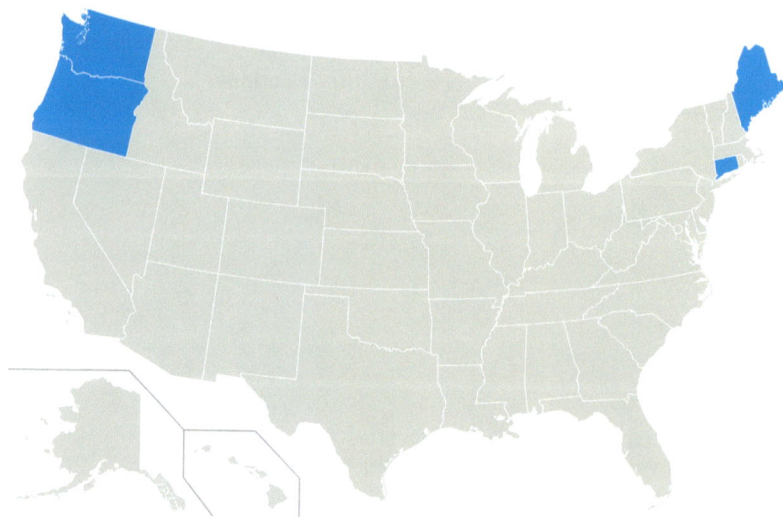

Cap Shape

Conical

Hymenium Shape

Gills

Hymenium Attachment

Adnate or Sinuate

Stipe Character

Bare Stipe

Psychoactive

Medium

Ecological Type

Saprotrophic

Spore Print Color

Purple-Brown

The conical cap of Psilocybe baeocystis ranges from 1.5 to 5.5 cm in diameter. The cap can be conic to convex, and expand to become planar when the mushroom reaches advanced age. The cap has a dark olive brown coloration that occasionally appears steel blue. The adnate to sinuate gills of this species range from gray to cinnamon brown in color. The bare stipe ranges from 50 to 70 mm in length and 2 to 3 mm in thickness and has a brownish gray surface color. The spore print color of P. baeocystis is purple-brown.

This species grows in solitude or scattered in small clusters on bark, wood chips, moss, decaying mulch, lawns, and pastures in temperate or subarctic climates from August through December.

It is psychoactive and contains the hallucinogens psilocybin and psilocin. This species has also been documented to produce psilocybin analogs baeocystin and norbaeocystin. It grows throughout the Pacific Northwestern United States and has also been found in Maine and Connecticut. The cap of P. baeocystis bruises blue when handled and has been observed to bruise very easily, even from the impact of rain drops. The hallucinogenic compound psilocin was first reported in this species.

References

[1] Borovička J, Oborník M, Stříbrný J, Noordeloos ME, Parra LA, Gryndler M, Phylogenetic and chemical studies in the potential psychotropic species complex of Psilocybe atrobrunnea with taxonomic and nomenclatural notes, Persoonia, 2015. Volume 34, Pages 1-9.

[2] Stamets, Paul (1996). Psilocybin Mushrooms of the World. Berkeley: Ten Speed Press. Pages 96-98. ISBN 0-9610798-0-0.

[3] https://www.shroomery.org/12496/Psilocybe-baeocystis

[4] Beug MW, Bigwood J, Quantitative analysis of psilocybin and psilocin in psilocybe baeocystis (Singer and Smith) by high-performance liquid chromatography and by thin-layer chromatography, Journal of Chromatography, 1981. Volume 207, Issue 3, Pages 379-85.

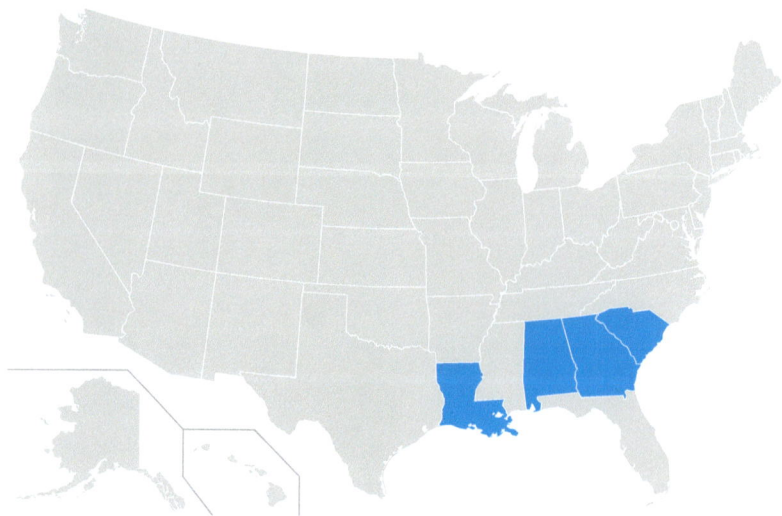

Cap Shape

Campanulate to convex

Hymenium Shape

Gills

Hymenium Attachment

Adnate or Sinuate

Stipe Character

Cortina

Psychoactive

High

Ecological Type

Saprotrophic

Spore Print Color

Purple-Brown

The cap of Psilocybe caerulescens ranges from 2 to 9 cm in diameter, and can appear obtusely campanulate to convex. Rarely, the mushroom's cap can become planar as the mushroom reaches advanced age. The cap's color is reddish brown to yellow brown. This species has a cap that bruises blue when handled. The adnate to sinuate gills range from whitish to yellowish gray in color with white edges. The stipe ranges from 40 to 120 mm in length and 2 to 10 mm in thickness, and has a brown surface and partial cortinate veil that disappears with age. Like the cap, the mushroom's stem bruises blue when handled. The spore print color is dark purple-brown, much like other species within the Psilocybe genus.

The species grows gregariously or in small clusters and is only very rarely found growing in solitude. P. caerulescens grows from late spring to summer in muddy clay soils, lawns, and deep woods where there is decaying wood. This species has been found in the United States, Mexico, Venezuela, and Brazil. Within the United States, it has been reported in Alabama, Georgia, and South Carolina. This species has been noted as potent.

References

[1] Borovička J, Oborník M, Stříbrný J, Noordeloos ME, Parra LA, Gryndler M, Phylogenetic and chemical studies in the potential psychotropic species complex of Psilocybe atrobrunnea with taxonomic and nomenclatural notes, Persoonia, 2015. Volume 34, Pages 1-9.

[2] Stamets, Paul (1996). Psilocybin Mushrooms of the World. Berkeley: Ten Speed Press. Pages 102-103. ISBN 0-9610798-0-0.

[3] https://mushroomobserver.org/name/show_name_description/1138

[4] https://www.shroomery.org/12498/Psilocybe-caerulescens

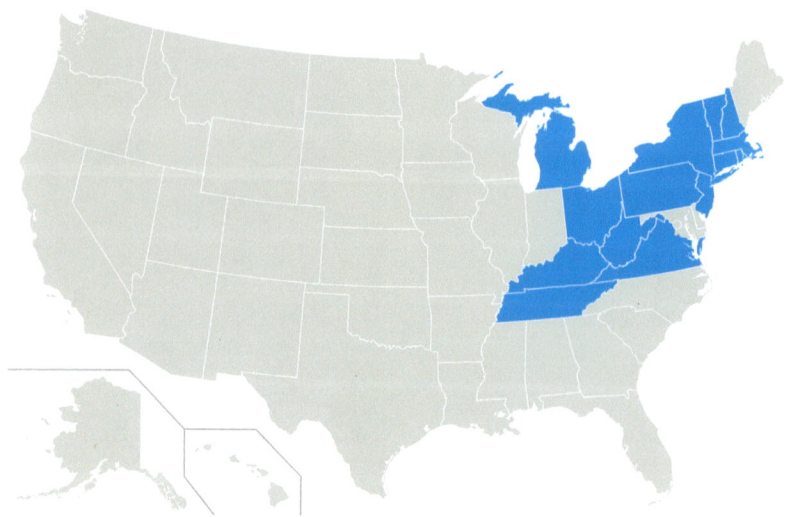

Cap Shape

Conic to Convex

Hymenium Shape

Gills

Hymenium Attachment

Adnate or Sinuate

Stipe Character

Bare Stipe

Psychoactive

Medium

Ecological Type

Saprotrophic

Spore Print Color

Purple

The obtusely conic to broadly convex cap of Psilocybe caerulipes ranges from 1 to 3.5 cm in diameter. The cap is conic when the mushroom is young, becoming conic-campanulate then broadly convex when the mushroom reaches advanced age. The cap's coloration is cinnamon brown to dingy brown and bruises blue when handled. The adnate to sinuate gills are narrow with white edges and a rusty cinnamon brown coloration. The stipe ranges from 30 to 60 mm in length and 2 to 3 mm in thickness. The stem's color is white with dingy brown lower regions. Like the cap, the stem bruises blue when handled. The spore print color of this species is purple-brown.

P. caerulipes grows in solitude to cespitose on hardwood debris, especially decaying birch, beech, and maple logs near rivers. The mushroom grows from summer to fall and is widely distributed in the Midwest and Eastern areas in the United States. The species is found from Ontario to North Carolina, and west to Michigan. The mushroom has also been reported in Mexico.

The species has been reported as moderately psychoactive. The common name of this mushroom is "blue foot." Although this species is distributed widely, it is rarely observed.

References

[1] Borovička J, Oborník M, Stříbrný J, Noordeloos ME, Parra LA, Gryndler M, Phylogenetic and chemical studies in the potential psychotropic species complex of Psilocybe atrobrunnea with taxonomic and nomenclatural notes, Persoonia, 2015. Volume 34, Pages 1-9.

[2] Stamets, Paul (1996). Psilocybin Mushrooms of the World. Berkeley: Ten Speed Press. Pages 104-105. ISBN 0-9610798-0-0.

[3] https://www.shroomery.org/12499/Psilocybe-caerulipes

[4] https://mushroomobserver.org/name/show_name/19919

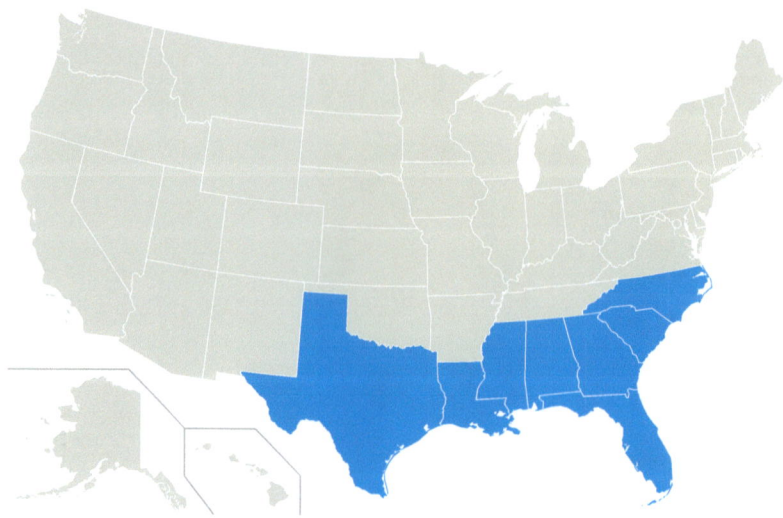

66

Cap Shape

Conic to Convex

Hymenium Shape

Gills

Hymenium Attachment

Adnate or Adnexed

Stipe Character

Ring

Psychoactive

Medium

Ecological Type

Saprotrophic

Spore Print Color

Purple

The conic to convex cap of Psilocybe cubensis ranges from 1.5 to 8 cm in diameter. The cap's color is a reddish cinnamon brown when the mushroom is young and becomes golden brown as the mushroom reaches advanced age. Like other psilocybin-containing mushrooms, the cap flesh bruises blue when handled. The gills of this species are adnate or adnexed and have a grayish to deep purple gray color. The stipe ranges from 40 to 150 mm in length and 5 to 15 mm in width. The stem's color is whitish or yellowish and bruises blue like the cap. The spore print of this species is colored deep purple or purple brown.

P. cubensis grows in a scattered fashion or gregariously on bovine dung, as well as horse or elephant dung, in the spring, summer, and fall. The species is found in Cuba, Mexico, Central America, South America, India, Thailand, Vietnam, Cambodia, Australia, and the United States. Within the United States, it is found in the southeastern states.

This species is psychoactive and has been noted to be moderately potent. Common names for this species are golden tops and cubies. P. cubensis is one of the most well-known mushroom species known for use in inducing psychedelic experiences.

References

[1] Borovička J, Oborník M, Stříbrný J, Noordeloos ME, Parra LA, Gryndler M, Phylogenetic and chemical studies in the potential psychotropic species complex of Psilocybe atrobrunnea with taxonomic and nomenclatural notes, Persoonia, 2015. Volume 34, Pages 1-9.

[2] Stamets, Paul (1996). Psilocybin Mushrooms of the World. Berkeley: Ten Speed Press. Pages 108-110. ISBN 0-9610798-0-0.

[3] https://www.shroomery.org/12500/Psilocybe-cubensis

[4] https://mushroomobserver.org/name/show_name/461

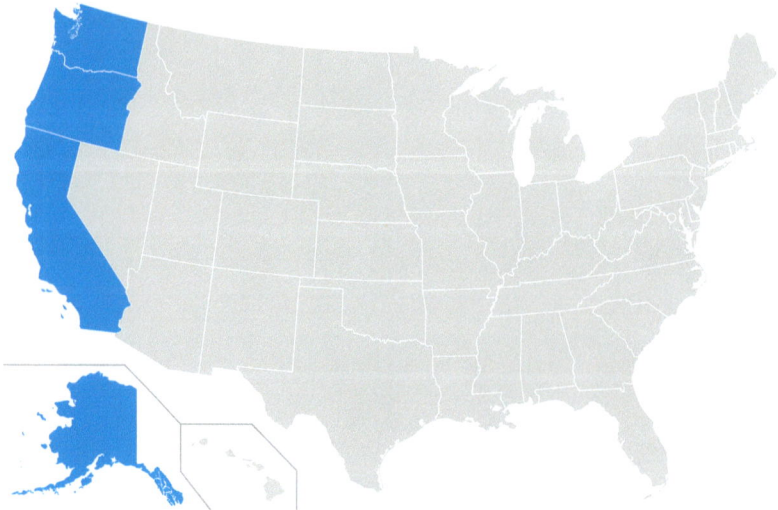

Cap Shape

Conic to Convex

Hymenium Shape

Gills

Hymenium Attachment

Adnate or Subdecurrent

Stipe Character

Cortina

Psychoactive

Medium

Ecological Type

Saprotrophic

Spore Print Color

Purple-Brown

The conic to convex cap of Psilocybe cyanescens ranges from 2 to 5 cm in diameter. The cap is conic to convex when the mushroom is young, and expands to broadly convex when the mushroom reaches advanced age. The cap's color is chestnut brown when the mushroom is new and becomes a caramel color with age. The gills of this species are adnate or subdecurrent and range from cinnamon brown to deep smoky brown in color. The stipe ranges from 20 to 80 mm in length and 2.5 to 5 mm in thickness. The stem's color is whitish and bruises blue when handled. There is a cortinate snow white partial veil that deteriorates into an annular zone. The spore print color of this species is dark purple-brown.

These mushrooms grow scattered to gregariously in woody debris and areas with rotten wood. In the United States, the species grows from fall to winter in the Pacific Northwest. P. cyanescens has been notably found on the west coast of the United States from San Francisco to the southern region of Alaska. Globally, it is found in the United Kingdom, Italy, Germany, Spain, and Sweden in temperate climates.

The species is psychoactive and has been reported to be moderately potent to highly potent. Common names for this species are cyans, blue halos, and wavy-capped Psilocybe.

References

[1] Borovička J, Oborník M, Stříbrný J, Noordeloos ME, Parra LA, Gryndler M, Phylogenetic and chemical studies in the potential psychotropic species complex of Psilocybe atrobrunnea with taxonomic and nomenclatural notes, Persoonia, 2015. Volume 34, Pages 1-9.

[2] Stamets, Paul (1996). Psilocybin Mushrooms of the World. Berkeley: Ten Speed Press. Pages 111-112. ISBN 0-9610798-0-0.

[3] https://www.shroomery.org/12501/Psilocybe-cyanescens

[4] https://mushroomobserver.org/name/show_name/1036

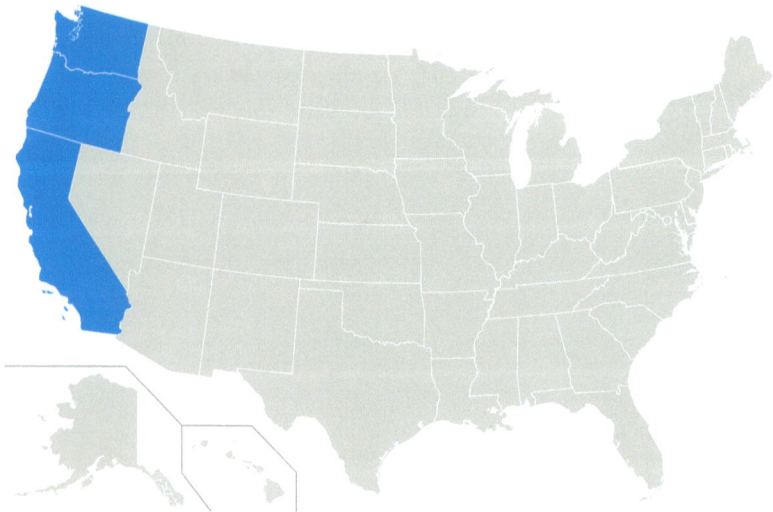

Cap Shape

Conic to Convex

Hymenium Shape

Gills

Hymenium Attachment

Adnate or Adnexed

Stipe Character

Cortina

Psychoactive

Low

Ecological Type

Saprotrophic

Spore Print Color

Purple-Brown

The conic to convex cap of Psilocybe cyanofibrillosa ranges from 1.4 to 3.5 cm in diameter. The deep chestnut brown cap is conic to broadly convex when the mushroom is young and becomes planar as the mushroom ages. The adnate to adnexed gills of this species can become subdecurrent when the mushroom reaches advanced age. The gills are purplish brown with white edges. The stem ranges from 30 to 70 mm in length and 2 to 4 mm in thickness and is yellow brown or brown in color. Like other Psilocybe mushrooms, the stipe bruises blue when handled. The spore print color of this species is purplish brown.

P. cyanofibrillosa grow gregariously to scattered along the west coast of the United States in flood plaines on river estuaries near the Pacific Ocean. The species is distributed from Northern California extending to the Canadian border and upward to British Columbia. Along its distribution route, this species is often found in rhododendron gardens, hence one of its common names, "rhododendron Psilocybe."

Like other mushrooms within the Psilocybe genus, this species is psychoactive and contains psilocybin and psilocin. This species has been observed to lose a large percentage of its potency when dried.

References

[1] Borovička J, Oborník M, Stříbrný J, Noordeloos ME, Parra LA, Gryndler M, Phylogenetic and chemical studies in the potential psychotropic species complex of Psilocybe atrobrunnea with taxonomic and nomenclatural notes, Persoonia, 2015. Volume 34, Pages 1-9.

[2] Stamets, Paul (1996). Psilocybin Mushrooms of the World. Berkeley: Ten Speed Press. Pages 113-114. ISBN 0-9610798-0-0.

[3] https://www.shroomery.org/12502/Psilocybe-cyanofibrillosa

[4] https://mushroomobserver.org/name/show_name/1117

PSILOCYBE FIMETARIA

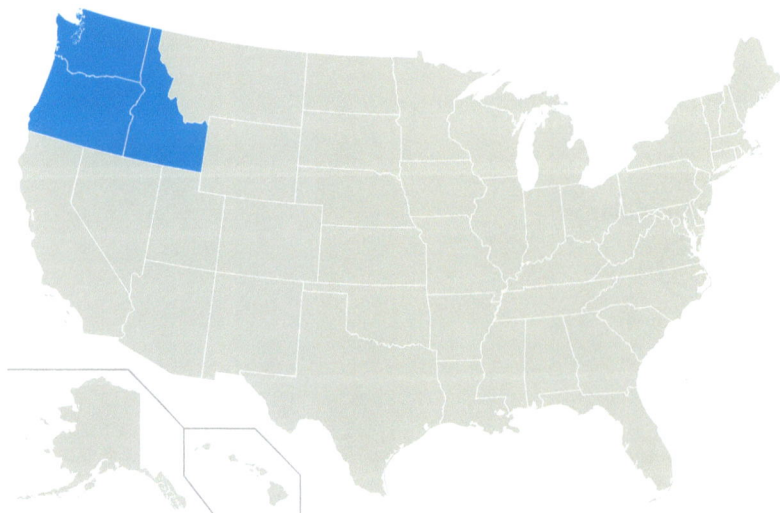

Cap Shape

Conic to Convex

Hymenium Shape

Gills

Hymenium Attachment

Adnate

Stipe Character

Cortina

Psychoactive

Medium

Ecological Type

Saprotrophic

Spore Print Color

Purple-Brown

The conic to convex cap of Psilocybe fimetaria ranges from 0.5 to 3.6 cm in diameter. The cap's coloration is reddish brown to honey brown. This species has adnate gills which are colored whitish gray when the mushroom is young and become dark purple-brown as the mushroom ages. The stipe ranges from 20 to 90 mm in length and 0.5 to 4 mm in thickness. The stem color is whitish when the mushroom is young, becoming reddish brown or honey brown as the mushroom reaches advanced age. The spore print color of this species is purple-brown. The cap flesh and stem bruise bluish when handled.

This species grows in solitude or gregariously on horse manure, grasses, and rich soils. This mushroom is distributed throughout the world, and is found in Canada, the Pacific Northwest region of the United States, Chile, Great Britain, and Europe. Within the United States, it has been located in Washington, Oregon, and Idaho.

Like other mushroom species within the Psilocybe genus, P. fimetaria is psychoactive. It has been reported to be moderately potent. In addition, it has been noted that P. fimetaria morphologically resembles related species Psilocybe subaeruginascens and P. stuntzii. P. fimetaria can be differentiated from these species because it grows readily on horse dung.

References

[1] Borovička J, Oborník M, Stříbrný J, Noordeloos ME, Parra LA, Gryndler M, Phylogenetic and chemical studies in the potential psychotropic species complex of Psilocybe atrobrunnea with taxonomic and nomenclatural notes, Persoonia, 2015. Volume 34, Pages 1-9.

[2] Stamets, Paul (1996). Psilocybin Mushrooms of the World. Berkeley: Ten Speed Press. Pages 113-114. ISBN 0-9610798-0-0.

[3] https://www.shroomery.org/12504/Psilocybe-fimetaria

[4] https://mushroomobserver.org/name/show_name/15174

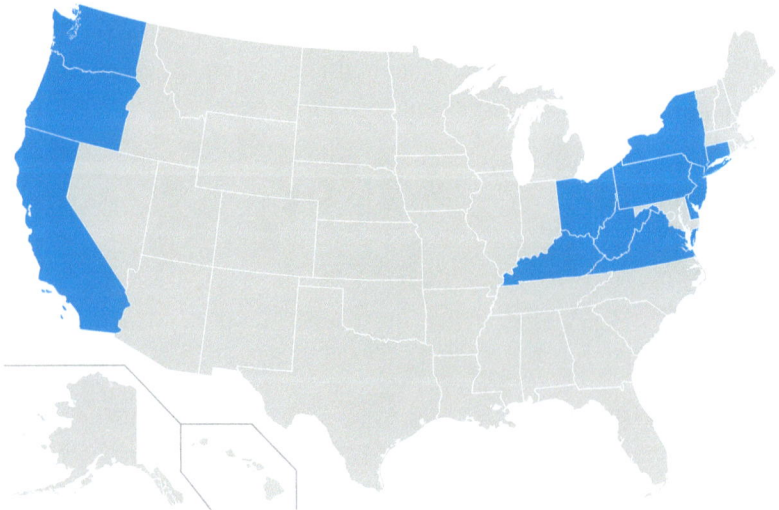

Cap Shape

Convex

Hymenium Shape

Gills

Hymenium Attachment

Adnate

Stipe Character

Cortina

Psychoactive

Low

Ecological Type

Saprotrophic

Spore Print Color

Purple-Brown

The convex to subumbonate cap of Psilocybe ovoideocystidiata is approximately 4.5 cm across in diameter. The mushroom's cap is a chestnut brown, orange-brown or yellow-brown to creamlike color. The species has been noted to sometimes sport irregular yellowish, brownish, or blue hues. Like other psilocybin-containing mushrooms, the cap flesh bruises blue when handled. The gills of this species range from whitish to rusty brown. The stipe ranges from 1.5 to 9 cm in length and 1 to 7 mm in thickness. The stem has been documented to be white with irregular zones of yellow, brown, or blue. The spore print is a dark purple brown color.

This species grows on wood chips and mulch in the Pacific Northwest region of the United States. Additionally, P. ovoideocystidiata is found throughout the Ohio River valley along rivers and streams and has been observed to grow in southern California. Globally, it has been documented in Switzerland and Germany.

This mushroom is psychoactive and contains the hallucinogens psilocybin and psilocin. Microscopically, this species has been observed to have distinct rhomboid spores. It was named partially for its microscopic ovoid pleurocystidia and cheilocystidia. Like other Psilocybe species, this mushroom has been observed to have a flour-like to spicy odor.

References

[1] https://www.shroomery.org/12508/Psilocybe-ovoideocystidiata

[2] https://www.researchgate.net/publication/235436020_The_Occurrence_Cultivation_and_Chemistry_of_Psilocybe_ovoideocystidiata_a_new_Bluing_Species_Agaricales_from_Ohio_Pennsylvania_and_West_VirginiaByAllen_John_W_Gartz_Jochen_Molter_Dan_and_Prakitsin_Sih

PSILOCYBE PELLICULOSA

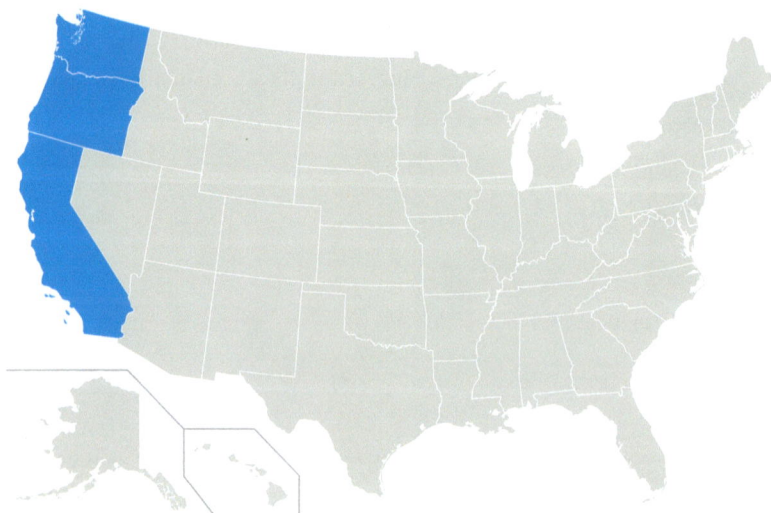

Cap Shape

Conical

Hymenium Shape

Gills

Hymenium Attachment

Adnate or Adnexed

Stipe Character

Bare

Psychoactive

Low

Ecological Type

Saprotrophic

Spore Print Color

Purple

Psilocybe pelliculosa has a conical cap that is obtusely conic when the mushroom is young, becoming conic-campanulate when the mushroom reaches advanced age. The cap is chestnut brown when the mushroom is fresh and turns dingy yellow when the mushroom is dry. The adnate to adnexed gills of this species are dull cinnamon brown. The stipe ranges from 60 to 80 mm in length and 1 to 2.5 mm in thickness. The stem is white to gray in color and may bruise blue green when handled. It has been noted that bluing or bruising sometimes does not occur with this species. The spore print color is purple.

P. pelliculosa grows gregariously, in scattered patches, and in solitude. The mushroom's habitat includes dead conifers, mulch, and lignin-rich soils. This species grows in the Pacific Northwest region of the United States and also grows in Northern California and British Columbia. One collection of this species has been found in Finland.

It is psychoactive and contains psilocybin, but not psilocin. P. pelliculosa grows from late October through December after cool, wet weather. This species is sometimes commonly referred to as the "conifer Psilocybe."

References

[1] Stamets, Paul (1996). Psilocybin Mushrooms of the World. Berkeley: Ten Speed Press. Pages 135-136. ISBN 0-9610798-0-0.

[2] https://www.shroomery.org/12509/Psilocybe-pelliculosa

PSILOCYBE SEMILANCEATA

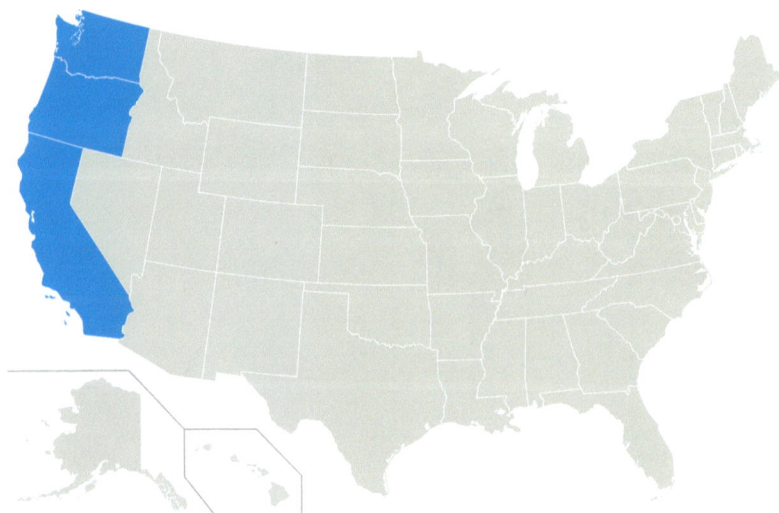

Cap Shape

Conical

Hymenium Shape

Gills

Hymenium Attachment

Adnexed

Stipe Character

Cortina

Psychoactive

High

Ecological Type

Saprotrophic

Spore Print Color

Purple-Brown

The cap of Psilocybe semilanceata is conic to obtusely conic-campanulate. The cap's coloration varies and ranges from dark chestnut brown when the mushroom is fresh to light tan or yellow when the mushroom is dry. The cap's diameter ranges from 0.5 to 2.5 cm. The gills of this species are adnexed and their color ranges from brown to purple-brown. The stem ranges from 40 to 100 mm in length and 0.75 to 3 mm in thickness. The stem is whitish brown and may bruise a blue color when drying. The spore print color of P. semilanceata is dark purple brown.

This species grows in scattered formations or gregariously in the fall season. Its habitat includes pastures, fields, lawns, and grasses. It can be found in the damp part of fields where sheep and cows are allowed to graze. Within the United States, the species grows throughout Northern California and along the coast to British Columbia. In the spring, it has been spotted in coastal Oregon and Washington. Globally, this species has been reported in Europe, South Africa, Chile, India, Australia, and Tasmania. P. semilanceata is psychoactive and contains the hallucinogens psilocybin and baeocystin. It has been observed to be a particularly potent species.

References

[1] https://www.sciencedirect.com/topics/neuroscience/psilocybe-semilanceata

[2] https://www.shroomery.org/12510/Psilocybe-semilanceata

[3] Stamets, Paul (1996). Psilocybin Mushrooms of the World. Berkeley: Ten Speed Press. Pages 142-145. ISBN 0-9610798-0-0.

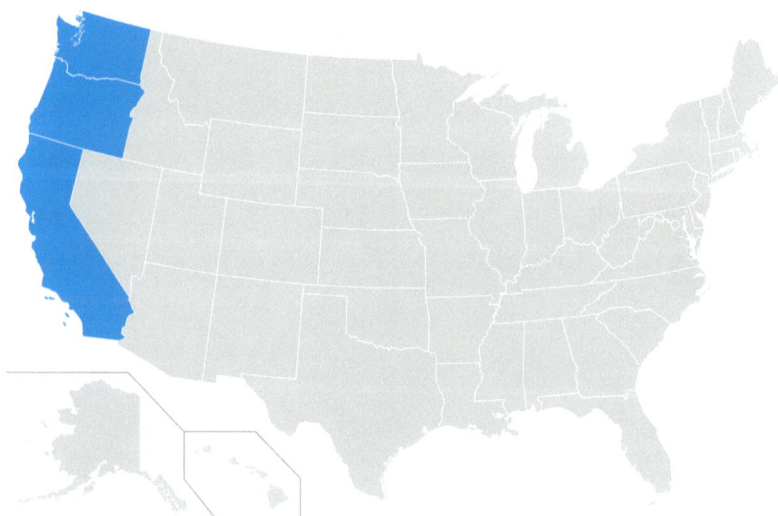

Cap Shape

Convex

Hymenium Shape

Gills

Hymenium Attachment

Adnate or Adnexed

Stipe Character

Bare

Psychoactive

Low

Ecological Type

Saprotrophic

Spore Print Color

Purple-Brown

The convex cap of Psilocybe stuntzii is conic when the mushroom is new, expanding to broadly convex as the mushroom reaches advanced age. The cap ranges from 1.5 to 5 cm in diameter and is dark chestnut brown in color. The cap flesh can have olive green tinges of color. The adnate to adnexed gills are brown to dark brown. The stem ranges from 30 to 60 mm in length and 2 to 4 mm in thickness. The stipe color is brownish yellow or pale yellow. The spore print color is purple-brown.

P. stuntzii grows gregariously or in small clusters in an array of habitats. The species grows on decaying conifers, wood chips, soils rich in decaying wooden debris and in cultivated gardens and worn paths. Within the United States, the species grows in Oregon, Washington, and upward along the Pacific coast toward British Columbia. The species is abundant among coastal areas in this region.

P. stuntzii is psychoactive and may contain varying amounts of the hallucinogens psilocybin, psilocin, and baeocystin. This species has been observed to be less potent by weight than other comparable mushrooms within the Psilocybe genus. It is notably similar in morphology to the poisonous species Galerina marginata; poisonings have occurred in foragers who have mistakenly consumed G. marginata when attempting to consume P. stuntzii.

References

[1] Stamets, Paul (1996). Psilocybin Mushrooms of the World. Berkeley: Ten Speed Press. Pages 151-152. ISBN 0-9610798-0-0.

[2] https://www.shroomery.org/12513/Psilocybe-stuntzii

[3] https://mushroomobserver.org/113977

GLOSSARY

A

Adnate Adnate mushroom gills are noticeably attached to the stem at a point slightly higher than the bottom of the gill. The gills are fused with the stem.

Adnexed Adnexed Mushroom gills are narrowly attached to the stem.

Aeruginascin A chemical compound found in the Inocybe aeruginascens mushroom species. It is structurally similar to a chemical found in toad venom called bufotenidine.

Agaric A mushroom type characterized by a fruiting body in an umbrella shape attached to a stem structure with gills underneath the cap.

Annular Zone A zone on the stem of a mushroom containing noticeable fibers where an annulus would be present, but is absent.

Apical The adjective for apex describing the tip of a pyramidal or rounded structure, such as a mushroom cap.

B

Baeocystin A minor chemical compound in most psilocybin mushrooms that is a derivative of psilocybin. Scientific studies have shown the chemical produces a hallucinogenic effect on humans.

Bare Stipe A mushroom stem void of any noticeable annulus or other common features.

C

Campanulate A cap structure that has the shape of a bell.

Cap The umbrella-like fruiting body resting atop a stem on agaric mushrooms. Spores are normally produced on the underside of the cap and contained in gills.

Cespitose Growth habit forming mat, clump, or dense turf like structures.

Cheilocystidia A type of cystidia found on the edge of a gill surface sometimes useful for identifying mushroom species.

Clade A group of organisms believed to be evolved from a common ancestor.

Convex A smoothly rounded cap shape. Most young mushrooms exhibit this type of cap shape.

Coprophilous Coprophilous mushrooms grow on animal dung. Spores of such species can be consumed by animals and excreted to then develop on the dung.

Cortinate Veil A type of partial veil produced by a fibrous tissue that appears weblike, especially on immature mushrooms.

Cystidia A large, oddly shaped cell located on the surface of a mushroom's gill. Such cells are distinguished from end cells and basidia and can often help an observer distinguish between species.

D

Decurrent Decurrent mushroom gills extend down the stem and attach well below the cap's point of connection.

F

Family The taxonomic group below

Order and above Genus.

Fibrillose Mushroom flesh made up of fibers or string-like material.

G

Genus The taxonomic group below Family and above Species.

Germ Pore A small pore in the wall of a fungal spore. The germ tube proceeds through the pore during germination. The location and color of the pore can be utilized in identification of particular species.

Gills Rows of a papery hymenophore located under the cap of agaric mushrooms. This anatomical feature assists fruiting mushrooms in spreading pores.

Gregarious Mushrooms growing in loose groups in the same area.

H

Hemispheric A cap structure that has the shape of a half circle, or half egg. Typically seen in young mushrooms, but also in some adult species.

Hymenium A layer of cells containing spore bearing cells such as basidia. In agaric mushrooms, such cells would be located on the surface of the gills.

L

Lamellae *See* Gills

M

Morphology A discipline of biology concerned with the physical structures of organism and their uses.

Mycology The branch of biology focused on the study of fungi.

N

Neurotoxins Toxins that destroy neurological tissue.

Norbaeocystin A chemical derivative of psilocybin found in certain mushroom species which has been found to be hallucinogenic.

P

Partial Veil A structure of tissue left hanging on the stem after the veil is torn during the maturation process of a mushroom's fruiting body.

Phenotypic Observable physical characteristics of an organism as a result of its genotype.

Phylogenetic A discipline of biology concerned with the evolutionary history and relationships among different species.

Pileipellis Outermost layer of tissue on the cap of an agaric mushroom. The color, texture, and other characteristics of this layer are helpful in the identification of different species.

Pileus *See* Cap

Planar A cap shape whereby the cap becomes an almost flat surface.

Polyphyletic A grouping of organisms that do not share a common evolutionary ancestor.

Psilocin Psychedelic chemical that is structurally similar to psilocybin and often found in so-called "magic mushrooms."

Psilocybin A psychedelic chemical naturally produced in some mushroom species, which acts as a hallucinogen when ingested by humans.

R

Ring Veil A structure on the stem of a

mushroom resembling a ring produced from the remnants of a partial veil.

S

Saprotrophic Having the characteristic of feeding on non-living, decaying biological matter.

Sinuate Sinuate gills have the same height for the majority of their length of travel toward the stem before exhibiting a shallower dip preceeding a re-extension and attachment to the stem.

Species The taxonomic group below Genus.

Spore The biological structure produced by fungi used for reproduction.

Spore Print The collection of spores onto a surface, often through the method of pressing a mushroom's cap onto the surface and allowing it to dry. The print will allow visualization of spore colors and the ability to view them with the aid of microscopy.

Squamules Small, scaly structures often found on the surface of mature mushroom caps.

Stem The stalk-like structure upon which a mushroom's cap rests.

Stipe *See* Stem.

Strophariaceae A family of fungi under the order Agaricales containing saprotrophic mushrooms with red-brown spore prints and apical germ pores.

Subdecurrent Subdecurrent gills run parallel to the cap for the majority of their length and suddenly dip toward the attachment point on the stem.

Subspecies The taxonomic grouping below Species and, in mycology, above variety.

Subumbonate A cap shape having a slight umbo near the center.

T

Taxonomy In biology, the science or discipline of classifying organisms into related groups.

U

Umbilicate A cap shape having a noticeable depression in the center near the area the stem attaches.

Umbonate A cap shape having a prominent umbo.

Umbo A noticeable bump in the center of a mushroom cap near the area where the stem attaches.

V

Veil A physical structure in the form of a thin membrane covering the cap of a mushroom and attaching to the stem. Usually seen in immature mushrooms and tearing away at maturation. Remnants of a veil are often seen in the annular zone of certain mushroom species.

Volva A bowl like membranous structure at the base of a mushroom that is a remnant of a veil.

www.ingramcontent.com/pod-product-compliance
Lightning Source LLC
Chambersburg PA
CBHW040126270326
41926CB00005B/89

9 781953 450999